# Springer Theses

Recognizing Outstanding Ph.D. Research

## Aims and Scope

The series "Springer Theses" brings together a selection of the very best Ph.D. theses from around the world and across the physical sciences. Nominated and endorsed by two recognized specialists, each published volume has been selected for its scientific excellence and the high impact of its contents for the pertinent field of research. For greater accessibility to non-specialists, the published versions include an extended introduction, as well as a foreword by the student's supervisor explaining the special relevance of the work for the field. As a whole, the series will provide a valuable resource both for newcomers to the research fields described, and for other scientists seeking detailed background information on special questions. Finally, it provides an accredited documentation of the valuable contributions made by today's younger generation of scientists.

## Theses are accepted into the series by invited nomination only and must fulfill all of the following criteria

- They must be written in good English.
- The topic should fall within the confines of Chemistry, Physics, Earth Sciences, Engineering and related interdisciplinary fields such as Materials, Nanoscience, Chemical Engineering, Complex Systems and Biophysics.
- The work reported in the thesis must represent a significant scientific advance.
- If the thesis includes previously published material, permission to reproduce this must be gained from the respective copyright holder.
- They must have been examined and passed during the 12 months prior to nomination.
- Each thesis should include a foreword by the supervisor outlining the significance of its content.
- The theses should have a clearly defined structure including an introduction accessible to scientists not expert in that particular field.

More information about this series at http://www.springer.com/series/8790

Dennis Wee Keong Neo

# Ultraprecision Machining of Hybrid Freeform Surfaces Using Multiple-Axis Diamond Turning

Doctoral Thesis accepted by
National University of Singapore, Singapore

Springer

*Author*
Dr. Dennis Wee Keong Neo
Department of Mechanical Engineering
National University of Singapore
Singapore
Singapore

*Supervisor*
Prof. A. Senthil Kumar
Department of Mechanical Engineering
National University of Singapore
Singapore
Singapore

ISSN 2190-5053 ISSN 2190-5061 (electronic)
Springer Theses
ISBN 978-981-10-4082-5 ISBN 978-981-10-4083-2 (eBook)
DOI 10.1007/978-981-10-4083-2

Library of Congress Control Number: 2017933944



Printed on acid-free paper

This Springer imprint is published by Springer Nature
The registered company is Springer Nature Singapore Pte Ltd.
The registered company address is: 152 Beach Road, #21-01/04 Gateway East, Singapore 189721, Singapore

# Supervisor's Foreword

The aims of this Ph.D. work were to achieve a seamless manufacturing of hybrid freeform surfaces using an advanced diamond turning process, understand the generation of surface accuracies (form errors) from the choice of cutting strategies, and simplify the complications of generating cutting path for such freeform surfaces. The breakthroughs behind the Ph.D. work were the development of novel multiple-axis diamond turning techniques to overcome the limitations of a conventional diamond turning process, an analytical model to optimize the generation of ultraprecise freeform surfaces, and an add-on tool path processor for CAD/CAM software solutions. This book is therefore aimed at researchers or postgraduate students with a strong machining background who is interested in making contributions either to the field of manufacturing ultraprecise freeform surfaces or in the field of optimization of the ultraprecision machining processes.

Dr. Dennis Neo proposed novel methodologies to overcome four manufacturing barriers. In Chap. 4, he explains how an integrated FTS/STS technique increases the working cutting depths without modifying the existing machine system. In Chap. 5, he shows how a four-axis Guilloche machining technique produces complex freeform surface, such as Fresnel lens array, in a single process. In Sect. 6.1, he analytically demonstrates how the surface accuracy is related to the spiral arc-length between the cutting points, in a surface analytical model of tool path generation—cutting linearization error—which defines as an error between the ideal profile and the actual tool path. Finally, in Chap. 7, he shows how a CAD/CAM-API system defines ultraprecise surface generation for non-mathematically derived freeform surfaces (i.e., CAD models) with the implementation of proposed methodologies.

All the results presented in this book can be extended to tackle other similar challenges in the manufacturing of complex freeform surfaces. In particular, Dr. Neo's findings in the multiple-axis machining processes which not only optimize the fabrication process of hybrid freeform surfaces, but also overcome the loss of symmetry in the machining of such surfaces. On top of that, his results also optimize the selection of cutting parameters and strategies from his surface analytical model. Furthermore, Dr. Neo's results in the add-on API tool can be extended to other commercial CAD software solutions for planning and conducting the manufacture of hybrid freeform surface and to make available open interfaces for the different adaption technologies. Thus, his results can be used to contribute the improvement of CAD/CAM supports for ultraprecision machining of complex hybrid freeform surfaces.

I would like to conclude this foreword with some highlights of Dr. Neo's work. These include their comprehensive analysis toward improving the tool path generation process and strategies, such as the addition of machining axes to compensate strategically the loss of non-rotating symmetrical axes (Sects. 4.1 and 5.1); the equation to compute analytically the maximum profile error as a function of spiral arc-length (Eq. 6.5); the add-on API as an alternative tool to provide a post-processing tool for tool trajectory of machining freeform surfaces (Sect. 7.1). Eventually, the latter result, as pointed out by Dr. Neo in this book, opens a new avenue for research based on the multiple-axis ultraprecision machining technologies for advanced freeform surfaces.

NUS, Singapore Prof. A. Senthil Kumar
September 2016

**Parts of this thesis have been published in the following journal articles/chapters from various books:**

i. *D.W.K. Neo*, A.S. Kumar and M. Rahman, '*CAx technologies for hybrid fast tool/slow slide servo diamond turning of freeform surface*', Proceedings of the Institution of Mechanical Engineers, Part B, J. Engineering Manufacture (2016) 230(8): 1465–1479

ii. *D.W.K. Neo*, A.S. Kumar and M. Rahman, '*Automated Guilloche machining technique for the fabrication of polygonal Fresnel lens array*', Precision Engineering, 2015, 41:55–62

iii. *D.W.K. Neo*, A.S. Kumar and M. Rahman, '*A novel surface analytical model for cutting linearization error in fast tool/slow slide servo diamond turning*', Precision Engineering, 2014, 38(4):849–860

iv. *W.K. Neo*, A.S. Kumar and M. Rahman, '*A novel method for layered tool path generation in the fast tool servo diamond turning of non-circular microstructural surfaces*', Proceedings of the Institution of Mechanical Engineers, Part B, J. Engineering Manufacture, 2013, 227(2):210–219, (*Top most downloaded paper*)

v. A. Senthil Kumar, M.R. Aravind Raghavendra, *W.K. Neo*, M. Rahman, '*Fast and Fine Tool Servo for Ultraprecision Machining in Comprehensive Materials Processing*' M. Rahman, Ed., Vol. 11: Advanced Machining Technologies, Elsevier Ltd., 2014, pp 61–88, 10.1016/B978-0-08-096532-1.01104-3

# Acknowledgements

Herein I would like to gratefully acknowledge all those people who have helped me to complete this thesis. First of all, I thank my supervisors from National University of Singapore, Associate Professor A. Senthil Kumar and Professor Mustafizur Rahman, for their excellent guidance, generous support, and precious encouragement throughout my four years' research. They not only provided me valuable knowledge regarding my research but also constantly shared their wisdoms and advices to improve my academic research and daily life.

I extend my deepest gratitude to my beloved wife, Duan Qingchuan, my eldest son, Cheng Hao, and my twin sons, Jun Tian and Jun Han, for their great care and long-lasting spiritual support during all these years.

Finally, I also want to express my appreciation to the staff of AML: Mr. Nelson Yeo Eng Huat, Mr. Neo Ken Soon, Mr. Tan Choon Huat, and Mr. Lim Soon Cheong for their time and support in operating the machines and instruments for my experiments. Also thanks to my lab-mates and friends: Dr. Asma Perveen, Dr. Minh Dang Nguyen, Dr. Aravind Raghavendra, Afzaal, Akshay, Huang Rui, and Malar for their academic help and inspiration. I also would like to thank Xmicro Solution Pte Ltd. loaning their Olympus LEXT OLS4000 3D measuring laser microscope for the measurements.

# Contents

**1 Introduction** . . . . . . . . . . 1
1.1 Hybrid Freeform Surfaces . . . . . . . . . . 2
1.2 Ultraprecision Machining of Hybrid Freeform Surfaces . . . . . . . . . . 5
1.3 Main Objectives of This Dissertation . . . . . . . . . . 6
1.4 Organization of This Dissertation . . . . . . . . . . 6
References . . . . . . . . . . 7

**2 Literature Review** . . . . . . . . . . 9
2.1 Multiple-Axis Ultraprecision Diamond Machining Techniques . . . . . . . . . . 9
2.1.1 Fast Tool Servo (FTS) . . . . . . . . . . 10
2.1.2 Slow Slide Servo (SSS) . . . . . . . . . . 14
2.1.3 Other Multiple-Axis Ultraprecision Machining Techniques . . . . . . . . . . 16
2.2 State-of-Art CAD/CAM/CAE Technologies . . . . . . . . . . 18
2.2.1 CAD/CAM Technology for Surface Generation . . . . . . . . . . 18
2.2.2 Surface Accuracy and Error Compensation Approaches . . . . 19
2.3 Concluding Remarks . . . . . . . . . . 22
References . . . . . . . . . . 23

**3 Initial Development of CAD/CAM Technologies** . . . . . . . . . . 27
3.1 CAD/CAM for Multiple-Axis Ultraprecision Machining Processes . . . . . . . . . . 28
3.1.1 Non-uniform Rational B-Spline Freeform Surfaces . . . . . . . . . . 28
3.1.2 CAD/CAM Interpolator for FTS/SSS Diamond Turning . . . . . . . . . . 29
3.2 API Methodology for CAD/CAM Software Development . . . . . . . . . . 31
3.2.1 Experimental Validation . . . . . . . . . . 34
3.3 Concluding Remarks . . . . . . . . . . 38
References . . . . . . . . . . 38

**4 Development of Hybrid FTS/SSS Diamond Turning** .......... 41
4.1 Principle of Layered Tool Trajectory .......... 41
4.2 Layered Tool Trajectory Control .......... 43
4.3 Experimental Validations .......... 47
4.4 Concluding Remarks .......... 50
References .......... 51

**5 Novel Surface Generation of Complex Hybrid Freeform Surfaces** .......... 53
5.1 Novel Surface Generation for Automated Guilloche Machining Technique .......... 53
5.2 Experimental Validations .......... 55
5.2.1 Critical Machining Parameters .......... 56
5.3 Concluding Remarks .......... 62
References .......... 64

**6 Development of Surface Analytical Model for Accurate Hybrid Freeform Surfaces** .......... 65
6.1 Surface Generation for FTS/SSS Diamond Turning .......... 66
6.1.1 Novel Surface Analytical Model .......... 66
6.1.2 Cutting Linearization Error .......... 68
6.2 Experimental Validation .......... 70
6.2.1 Evaluation of Critical Machining Parameters .......... 70
6.2.2 Cutting Experiments and Results .......... 77
6.3 Concluding Remarks .......... 80
References .......... 81

**7 Integration and Implementation** .......... 83
7.1 Integrated CAD/CAM System .......... 83
7.1.1 Integrated Sub-system for AGMT Process .......... 84
7.1.2 Integrated Sub-system for Diamond Turning Process .......... 84
7.1.3 Optimization of Tool Geometry .......... 86
7.1.4 Geometrical Splitting of Hybrid Freeform Surface .......... 88
7.2 Case Study 1: Hexagonal Fresnel Lens Array Using AGMT Process .......... 89
7.2.1 Experimental Validations .......... 89
7.3 Case Study 2: Multiple-Compound Eye Surface Design-B .......... 97
7.3.1 Experimental Validations .......... 97
7.3.2 Cutting Experiments and Results .......... 102
7.4 Concluding Remarks .......... 104
References .......... 105

**8 Conclusions and Recommended Future Works** .......... 107
8.1 Major Contributions .......... 107
8.2 Recommended Future Works .......... 108

**Curriculum Vitae** .......... 111

# Acronyms

| | |
|---|---|
| AGMT | Automated Guilloche machining technique |
| API | Application programming interface |
| CAE | Computer-aided engineering |
| FFT | Fast Fourier transformation |
| FTS | Fast tool servo |
| HCAA | Hybrid constant-arc and constant-angle |
| HT | Hilbert transformation |
| MLA | Microlens array |
| NURBS | Non-uniform rational B-splines |
| SAM | Surface analytical model |
| SCD | Single-crystal diamond |
| SSS | Slow slide servo |
| SWG | Sinusoidal wave grid |

# Symbols

| | |
|---|---|
| $r$ | Outer radius of workpiece |
| $f_r$ | Radial feed per radian |
| $N_t$ | Total number of spiral rotations to reach the center from the outer radius |
| $\rho$ | Radial position of the tool from the center of workpiece |
| $\theta$ | C-axis of spindle or angular position of a spiral point |
| $x$ | X-axis which controls the radial movement toward the spindle center and is also perpendicular to spindle axis (Z-axis) |
| $Z$ | Z-axis which controls the axial movement along the spindle axis |
| $W$ | W-axis of the FTS stroke which controls the feed direction into the workpiece surface and is parallel to Z-axis |
| $i$ | $i$th angular position of workpiece or spindle |
| $W_{max}$ | Maximum stroke zone of FTS |
| $\Delta\theta$ | Constant-angle |
| $N_p$ | Number of control points per rotation |
| $E_z$ | Overcut depth of machined surface |
| $\beta$ | Surface slope along the feed direction |
| $P_i^*$ | Exit/reentry point |
| $r_t$ | Tool nose radius of diamond tool |
| $W^*$ | Effective stroke length of FTS |
| $W_c$ | Compensated FTS stroke length of tool trajectory |
| $Zb_{max}$ | Maximum Z-axis boundary |
| $Zr$ | Z-axis retraction |
| $Z_R$ $(Pi^*)$ | Minimum value for intersection point of surface and cylindrical region within a circumscribed radius $\rho(P_i^*)$ |
| $Cx$ | $X$-axis coordinates of the cutting point $P$ in AGMT |
| $Cy$ | $Y$-axis coordinates of the cutting point $P$ in AGMT |
| $t$ | Rotational position for the workpiece or spindle in AGMT |
| $Xc$ | $X$-axis of the center coordinates for an arc of the circular Fresnel lens in the AGMT |

| | |
|---|---|
| $Yc$ | $Y$-axis of the center coordinates for an arc of the circular Fresnel lens in the AGMT |
| $N_s$ | Number of sides in a polygon |
| $P$ | Cutting point or spiral point |
| $r_c$ | Arc-radius of circular tool trajectory of AGMT |
| $r_{lens}$ | Lens radius of a microlens |
| $r_p$ | Radius of polygonal tool trajectory with respect to $\rho$ |
| $C$ | Lens curvature of a microlens |
| $T$ | Remainder value of $t$ divided by 360° |
| $T_p$ | Angle between apothem of polygon and the Guilloche tool trajectory point |
| $\Delta r$ | Radii difference between the lens curvature $r_{lens}$ and tool nose radius $r_t$ |
| $\delta$ | Angular position of tool profiles with respect to the center of lens curvature at point $O$ in Fig. 5.6 |
| $\Delta\delta$ | Angle between two tool profiles along radial feed direction in Fig. 5.6 |
| $\Delta d$ | Distance $AB$ in Fig. 5.6 |
| $a_\delta$ | Apothem of the triangle $AOB$ in Fig. 5.6 |
| $a_f$ | Apothem of hexagonal Fresnel lens |
| $d_\delta$ | Euclidean distance from the midpoint of $AB$ to the tip of cusp in Fig. 5.6 |
| $d_f$ | Relief depth of Fresnel zone plate |
| $\Delta\rho$ | Feed rate |
| $\Delta\rho_{cr}$ | Critical feed rate |
| $E_\rho$ | Cutting residual error |
| $h_{err}$ | Sagitta errors |
| $h_{tol}$ | Sagitta of the chord which represents the maximum permissible profile error |
| $S$ | Arc-length from the center of the workpiece to a cutting point $P$ |
| $\Delta S$ | Constant-arc |
| $S_t$ | Arc-length for the entire spiral tool trajectory |
| $\theta_t$ | Total angular of spiral rotations to reach the center from the outer radius |
| $PV_{err}$ | Peak-to-valley errors |
| $\delta_{err}$ | Local $pv_{err}$ |
| $\lambda$ | Wavelength of SWG surface |
| $\phi$ | Slope of tool trajectory in the cutting direction |
| $\partial Z_{\max}$ | Maximum deviation between two corresponding cutting points |
| $A_{SWG}$ | Amplitude of SWG surface |
| $PV_{tol}$ | Profile accuracy tolerance |

# Summary

Hybrid freeform surfaces have been emerging to bring novel functionalities and applications in the optics industries. Hybrid freeform surfaces are designed with an integration of multiple freeform surfaces to increase their optical performance and provide new optical functions. Over the last several decades, ultraprecision machining technology has been evolving to fabricate most freeform optical surfaces that could not have been previously machined or machining them was expensive. Some of the known machining technologies to machine freeform optics use micromilling, raster flycutting, fast tool servo (FTS), and slow slide servo (SSS).

Micromilling requires overcoming inherent static and dynamic limitations in the ultraprecision machine system, and in this process, material removal rate is much lower than the turning process. Raster flycutting has several shortcomings to overcome such as relatively long setup time, difficult setup, and restriction of tool swing diameter. On the other hand, FTS and SSS diamond turning processes have the highest material removal rates as compared to other processes and therefore are widely used by many researchers and industries. However, only few studies have been conducted for the optimization of FTS and SSS processes to fabricate hybrid freeform surfaces. Based on the above facts, the optimization of FTS and SSS processes has been carried out in this dissertation.

In this dissertation, comprehensive studies have been conducted for the seamless manufacturing of hybrid freeform surfaces with good surface quality and accuracy. This dissertation consists of four major studies to contribute the optimization of manufacturing hybrid freeform surfaces.

Hybrid freeform surfaces with larger depths are difficult to machine using diamond turning. Hence, a hybrid fast tool/slow slide servo (FTS/SSS) diamond turning was developed by incorporating both FTS and SSS techniques to optimize the fabrication process of hybrid freeform surfaces. This technique addresses the limited range of FTS travel length and the low bandwidth in the SSS system.

Hybrid freeform surfaces in general have a loss of symmetry due to their complexity in the curvatures. It is necessary to increase the number of machining axes for moving a tool to produce such surfaces. Hence, a novel automated Guilloche machining technique with 4-axis CNC system to fabricate a complex hybrid freeform surface, such as a polygonal Fresnel lens array, has been developed to address the difficulties of fabricating such surfaces in a single setup.

A novel surface analytical model has been derived to pre-evaluate the accuracy of the machined freeform surface. The model evaluates the cutting linearization errors along the spiral tool trajectory of fast tool/slow slide servo diamond turning process and also optimizes the number of cutting points for achieving the targeted accuracy.

Most of commercial CAD/CAM software solutions for freeform surfaces are only suitable for Cartesian coordinate system, which do not support the FTS/SSS turning (polar/cylindrical coordinates) and also have a larger resolution range of 10 nm. A specialized CAM system is necessary to support FTS/SSS turning and have a better resolution range. Thus, a comprehensive, integrated CAD/CAM software solution for multiple-axis diamond turning process has also been developed for planning and conducting the manufacture of hybrid freeform surfaces.

In this dissertation, a comprehensive and integrated CAD/CAM software solution with the methodologies from the above studies has been developed and implemented. Thus, a seamless multiple-axis ultraprecision machining technology for hybrid freeform surface with good surface quality and accuracy has been successfully developed, implemented, and validated in this study.

# Chapter 1
# Introduction

Freeform optical surfaces are widely used to reduce wavefront error and sizes as compared to rotational surfaces. Ultraprecision machining techniques such as diamond turning with fast tool/slow slide servo (FTS/SSS) and diamond micromilling techniques are widely employed for machining freeform optical surfaces with ultraprecision accuracy and excellent surface quality.

Over the last several decades, these ultraprecision machining techniques are evolving to meet the demands of ultraprecision accuracy and excellent surface quality of freeform optical surfaces. This evolution in-turn marks the tipping point for the evolution of novel optical designs. These new evolutions have not been fully explored to unleash the hidden potential of freeform optical surfaces. This new field also brings us many new challenges in designing, machining and testing.

This chapter reports the current trends in ultraprecision machining techniques employed for generating hybrid freeform surfaces. Section 1.1 discusses the new era of hybrid freeform surfaces with their functionalities and applications. Section 1.2 highlights a great deal of challenges and machining barriers in this research area to be discussed for optimizing developments of these ultraprecision machining techniques to new higher levels. Section 1.3 gives a list of objectives for contributing the motivation to complete this dissertation. Lastly, Sect. 1.4 presents the organization of this dissertation, which summarizes several areas of improvements in the manufacturing of hybrid freeform surfaces.

---

Some content of this chapter has been reproduced with permission from [11]



D.W.K. Neo, *Ultraprecision Machining of Hybrid Freeform Surfaces Using Multiple-Axis Diamond Turning*, Springer Theses,
DOI 10.1007/978-981-10-4083-2_1

## 1.1 Hybrid Freeform Surfaces

There is a growing trend of designing freeform optical surfaces with hybrid freeform surfaces [1–7] for non-imaging devices such as solar concentrators and collimators to increase their optical performance, and imaging devices to achieve special imaging effects [7]. Simultaneous multiple surface (SMS) [1–4, 6] is one of the latest designing techniques, which can design N rotationally-symmetric surfaces that, by definition, form sharp images of Nth one-parameter subsets of rays allowing the control of extended sources. This design strategy consists of finding the best configuration of these subsets of rays in phase-space, one that ensures that image-quality specifications will be met by all rays. This gives better control of exit aperture shape without efficiency loss and increases tolerances to source displacement. It would be a challenging task to produce this new generation of freeform surfaces, as illustrated in Figs. 1.1, 1.2, 1.3, 1.4, 1.5 and 1.6, by conventional diamond machining techniques.

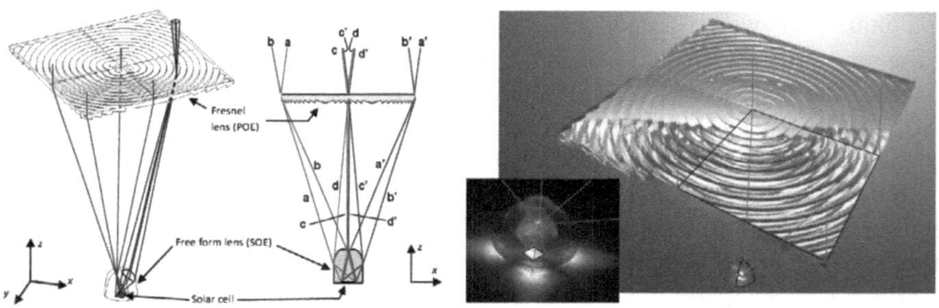


**Fig. 1.1** Four-fold Fresnel-Kohler (FK) concentrator [2] schematic diagram (*left*); rendered views (*right*)

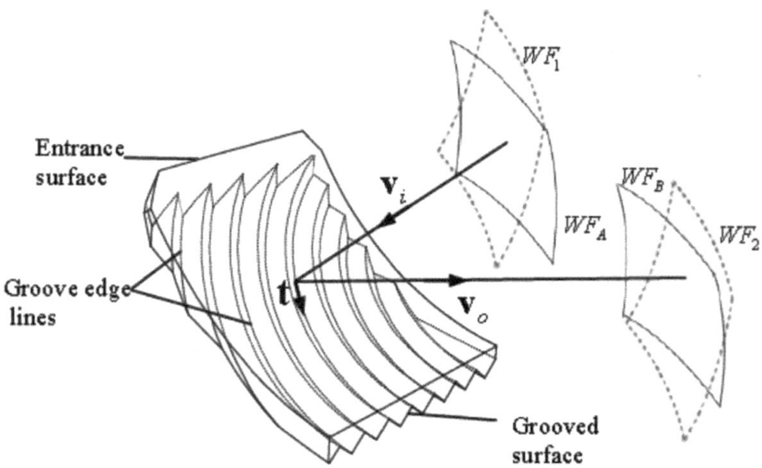


**Fig. 1.2** Freeform thin dielectric sheet as a TIR reflector [3]

Fig. 1.3 Metal-less TIR RXI collimator [4]

Fig. 1.4 Freeform reflector to eliminate the driver's blind spot effect [5]

Thanks to the state-of-art technologies, these hybrid freeform surfaces can be easily manufactured by multiple-axis diamond machining techniques. Basically, an increasing complexity is often associated with a loss of symmetry of the surface. With an increase in the number of degrees of freedom needed for moving a tool to produce a surface, the number of controllable machine axes will be increased. The applications and principles of these multiple-axis ultraprecision machining processes for the manufacturing of hybrid freeform surfaces are discussed in the next section.

Fig. 1.5 Ultra-short throw projector by LPI [6]

Fig. 1.6 Freeform mirror was used for special movie effect in an Oscar-nominated film, "Sleepless in New York" [7]

## 1.2 Ultraprecision Machining of Hybrid Freeform Surfaces

Over the past several decades, the diamond machining are evolving into ultraprecision machining techniques [8] which are capable of performing the machining of these freeform surfaces. Four common ultraprecision machining techniques are fast tool servo (FTS), slow slide servo (SSS), raster flycutting and micro-milling. These techniques have exhibited their machining capabilities to produce complex surfaces like lens arrays, bi-conics NURBS defined freeform surfaces, etc.

Figure 1.7 shows a process chain evaluating the feasibility of fabrication methods for freeform surfaces from the design to metrology [9]. This process chain allows a computer-aided manufacturing (CAM) software to generate and modify/correct the tool trajectory with the compensation of surface form error from the metrology process.

FTS diamond turning has been widely employed for fabricating the non-rotational symmetrical surfaces due to its high resolution and bandwidth [10]. Although SSS technique has a longer travel length up to several millimeters, its limited bandwidth restricts the speed of Z-axis (in the tool trajectory) for machining a complex freeform surface [11]. When raster flycutting is employed, there are several shortcomings to overcome such as relatively long and difficult setup and restriction of tool swing diameter [11]. Lastly, micro milling method requires overcoming of inherent static and dynamic limitations in the ultra-precision machine system and material removal rate is much lower than the turning process [11]. Therefore, FTS and SSS diamond turning is often employed for machining freeform surfaces [11]. In order to machine a hybrid freeform surface with large sag height, we need to have an ultraprecision machine which has the capability to machine at a larger depth and a system to generate accurate NC codes quickly and easily.

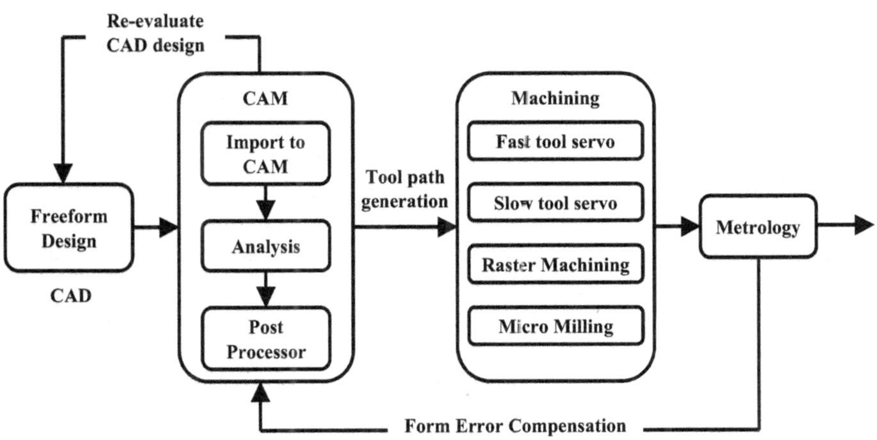


**Fig. 1.7** Process chain for the fabrication of freeform surfaces [9]

## 1.3 Main Objectives of This Dissertation

This dissertation aims to achieve a seamless manufacturing of hybrid freeform surface with good surface quality and accuracy using the diamond turning process. The main objectives are to be fulfilled, as follows:

i. To address the limited stroke distances and bandwidths for the FTS and SSS technologies in generating hybrid freeform surface with large curvature depths;
ii. To address the difficulties in machining complex hybrid freeform surfaces which cannot be machined by FTS and SSS processes;
iii. To conduct a process optimization of machining hybrid freeform surfaces in generating accurate tool trajectory control points with ultraprecise surface accuracy;
iv. To address the need for an alternative and economical option of specialized CAD/CAM system in generating accurate complex hybrid freeform surfaces for FTS/SSS and other multiple-axis diamond turning processes.

## 1.4 Organization of This Dissertation

This dissertation discusses several areas of improvement for diamond turning of hybrid freeform surfaces in the following chapters:

- Chapter 2 presents a literature survey which has been conducted on the studies of the manufacturing of hybrid freeform surfaces. A list of literature loopholes are also highlighted for this dissertation.
- Chapter 3 introduces an alternative method of surface generation for FTS/SSS diamond turning of freeform surface directly from computer-aided design (CAD) software with an integration of application programming interface (API).
- Chapter 4 discusses the hybrid FTS/SSS process with a novel tool trajectory generation technique by means of several layers of tool trajectory to overcome the short FTS travel length and low bandwidth of SSS system.
- Chapter 5 discusses a novel automated Guilloche machining technique (AGMT), offering capabilities to produce of complex freeform surfaces such Fresnel lens which cannot be machined by FTS/SSS diamond turning.
- Chapter 6 discusses a novel surface analytical model which evaluates the cutting linearization errors in the FTS/SSS diamond turning process. The accuracy of machined freeform surface can be pre-evaluated with the derived novel surface analytical model before machining stage.
- Chapter 7 discusses the integration and implementation of developed methodologies in the developed CAD/CAM system. This integration shall plan and

conduct the manufacture of hybrid freeform surface within the multiple-axis diamond turning process.

- Lastly, Chap. 8 highlights the conclusions of this dissertation and recommends some future works to be done.

## References


1. Muñoz F, Benítez P, Miñano JC. High-order aspherics: the SMS nonimaging design method applied to imaging optics. Proc SPIE 2008;7061:70610G–9.
2. Benítez P, Miñano JC, Zamora P, Mohedano R, Cvetkovic A, Buljan M, Chaves J, Hernández M. High performance Fresnel-based photovoltaic concentrator. Opt Express 2010;18(S1):A25–40.
3. Grabovičkić D, Benítez P, Miñano JC. Free-form V-groove reflector design with the SMS method in three dimensions. Opt Express. 2011;19(S4):A747–56.
4. Grabovičkić D, Benítez P, Miñano JC. TIR RXI collimator. Opt Express. 2012;20(S1): A51–61.
5. Wallace J. Reflective optics: free-form reflector eliminates driver's blind spot. Laser Focus World 2008;44(10).
6. Advanced imaging. http://www.lpi-llc.com/Advanced.php.
7. Salem, Sleepless in New York—Motion Picture with free-form optic made by Kugler, News article, May 2014, Kugler.
8. Riemer O. Advances in ultra precision manufacturing. In: International symposium of the Japan Society for Precision Engineering 2011 (ISUPEN 2011).
9. Tohme Y. Trends in ultra precision machining of freeform optical, optical fabrication and testing, OSA technical digest (CD) (Optical Society of America, 2008), paper OThC6.
10. Davis GE, Roblee JW, Hedges AR. Comparison of freeform manufacturing techniques in the production of monolithic lens arrays. Proc SPIE 2009;742605-1.
11. Neo WK, Kumar AS, Rahman M. A novel method for layered tool path generation in the fast tool servo diamond turning of non-circular microstructural surfaces. In: Proc Inst Mech Eng Part B, J Eng Manuf 2013;227(2):210–19.

# Chapter 2
# Literature Review

In this chapter, a literature survey on the manufacturing processes of hybrid freeform surfaces is presented. Section 2.1 discusses the main principles and the limitations of FTS/SSS diamond turning and other multiple-axis diamond machining techniques. Section 2.2 covers the existing CAD/CAM/CAE technologies employed for the manufacturing of hybrid freeform surfaces, and discusses the needs for the surface generation methodologies to produce an accurate hybrid freeform surface. Lastly, Sect. 2.3 presents the concluding remarks that lead to this dissertation.

## 2.1 Multiple-Axis Ultraprecision Diamond Machining Techniques

Freeform surfaces play the key role in development of complex optical devices widely used in telecommunication, medical imaging, and surveillance systems. Freeform surfaces also allow freedom for the optics designer to design products with functional, aesthetic, and ergonomic surfaces. Ultraprecision multi-axis freeform machining techniques are often employed for manufacturing freeform surfaces with high degree of accuracy and precision. Diamond turning is one of the ultraprecision machining techniques, which has the advantages like high accuracy and high efficiency. It is often coupled with unique technique known as fast tool/slow slide servo (FTS/SSS) technologies (as shown in Fig. 2.1) for machining a freeform surface with high degree of complexity due to its high resolution and bandwidth. FTS diamond turning integrates a high bandwidth servo unit in an additional W-axis (or superimposed Z-axis) with the existing three axes (X, Z and C-axis) in ultraprecision turning machine [1]. Unlike FTS method, SSS diamond turning uses the existing Z-axis to oscillate the tool. Some of the freeform optical surfaces manufactured by FTS and SSS diamond turning processes are illustrated in Figs. 2.2 and 2.3, respectively.



D.W.K. Neo, *Ultraprecision Machining of Hybrid Freeform Surfaces Using Multiple-Axis Diamond Turning*, Springer Theses,
DOI 10.1007/978-981-10-4083-2_2

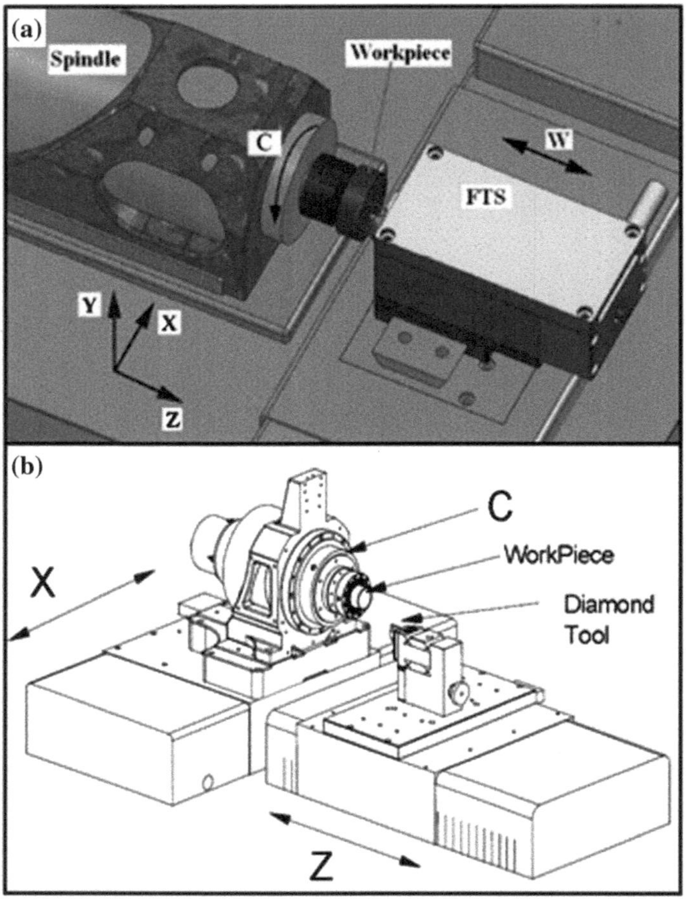


**Fig. 2.1** Configurations of ultraprecision lathe machines; **a** fast tool servo and **b** slow slide servo

## *2.1.1 Fast Tool Servo (FTS)*

Fast Tool Servo (FTS) technology plays an important role in machining complex freeform surfaces for the modern optics industry. Hence, FTS diamond turning has been widely employed for fabricating the non-rotational symmetrical surfaces due to its high resolution and bandwidth [1, 7]. Some of the works on FTS, dated back as early as 1980s, Meinel et al. [8] has successfully produced phase corrector plates for wavefront correction, and Luttrell [9] was able to fabricate off-axis conic surfaces and tilted flats with the FTS.

Unfortunately, most of FTS systems have limited travel of less than 1 mm, which makes it inappropriate for machining freeform surfaces with sag height greater than 1 mm [10]. Hence, there are several works to address this setback by increasing the FTS travel length to fulfill the sag height requirement. Several ways

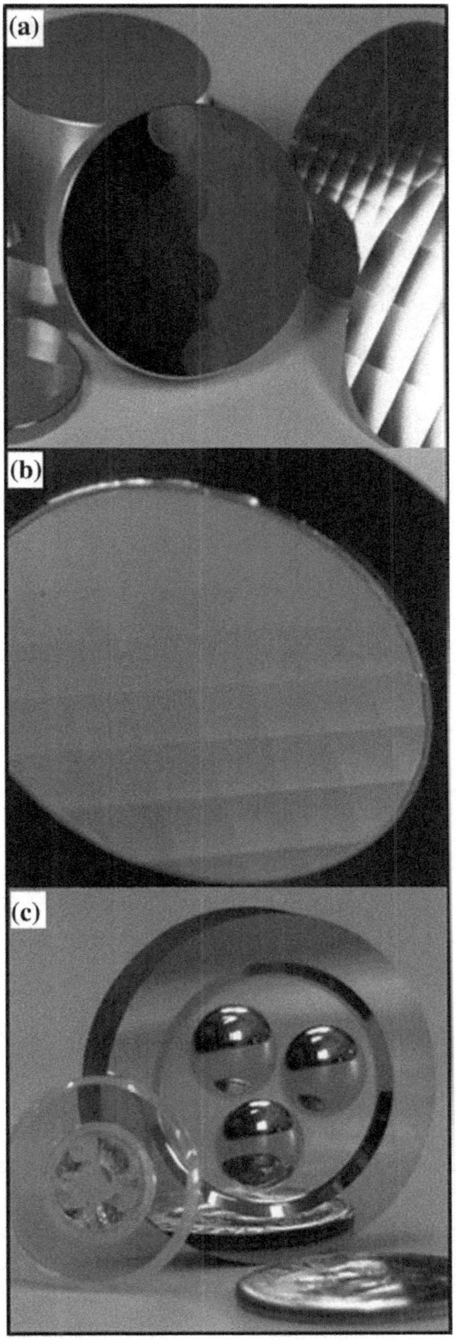


**Fig. 2.2** Freeform optical surfaces by FTS process [2]; **a** faceted mirrors, **b** faceted lenses, and **c** aspheric lens array

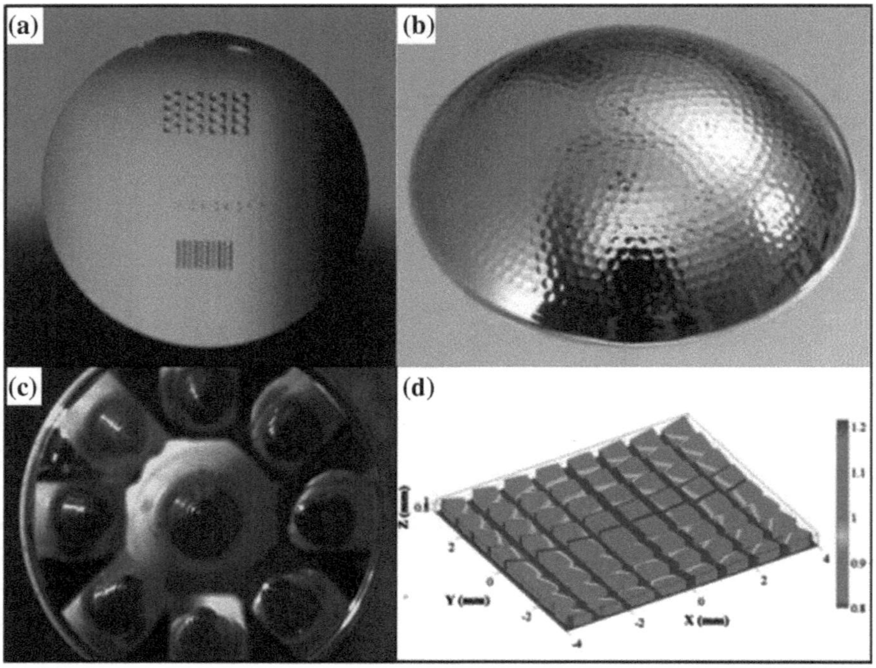


**Fig. 2.3** Freeform optical surfaces by SSS process [3–6]; **a** micro Alvarez lens array, **b** artificial compound eye, **c** freeform prismatic lens and **d** 8 × 8 freeform microlens array

of extending the travel of FTS are by using rotary FTS [11] and designing flexure of higher displacement amplification mechanism incorporated with voice coil and/or piezoelectric actuators [12, 13]. Figure 2.4 shows a rotary FTS design that has a travel up to 10 mm with a frequency of 50 Hz. However, it has been reported that there is a large tool position error due to a higher harmonic frequency error during the cutting process.

A long-stroke FTS (LFTS) can also be designed by incorporating with displacement amplification mechanism composed of several levers and hinges, which can be driven by a piezoelectric actuator [12] (Fig. 2.5) or a voice coil [13] (Fig. 2.6).

Although piezoelectric (PZT) actuator may have better resolutions of positioning, it has a resonance due to its low resonant frequency, and a vibration would also induce due to its low stiffness. These would deteriorate the machined surface quality. On the other hand, voice coil actuators are hysteresis-free which can achieve an almost linear current versus force relationship for smaller travels. This is a merit over the PZT actuators which employs a charge control to avoid hysteresis. However, the voice coil FTS usually has a lower frequency. These limitations can be overcome by integrating both PZT and voice coil actuators [14], as illustrated in Fig. 2.7. In this hybrid FTS, a voice coil actuator (VCM) drives a macro-range travel whereas a PZT actuator drive a fine micro-range travel.

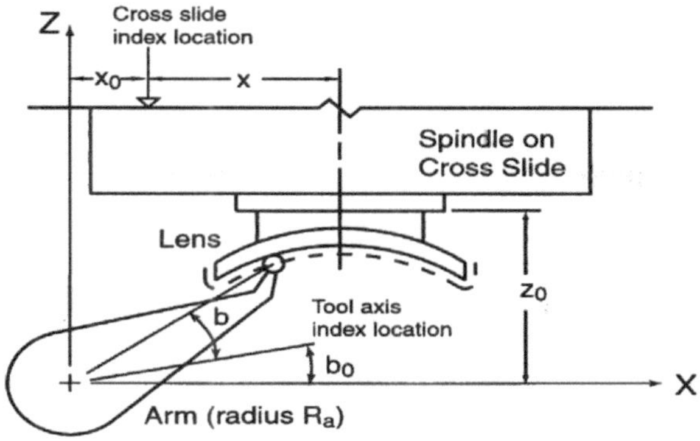


**Fig. 2.4** Rotary FTS [11]

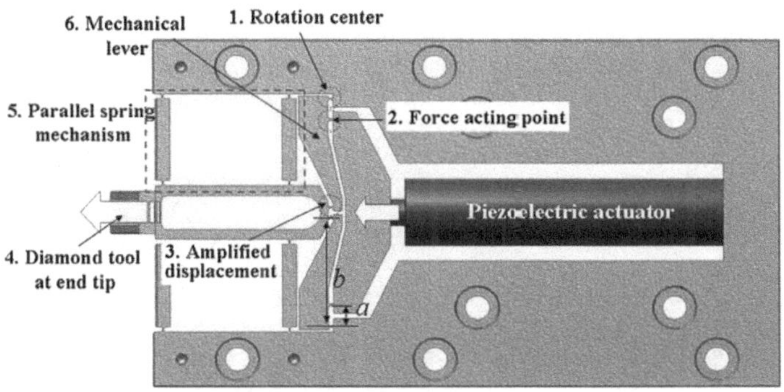


**Fig. 2.5** LFTS with PZT actuator [12]

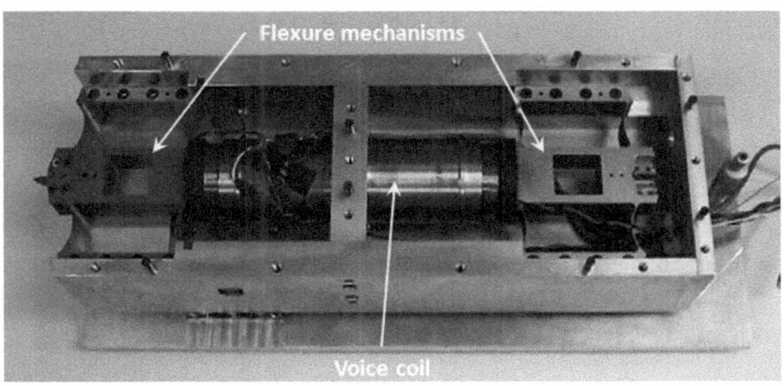


**Fig. 2.6** LFTS with voice coil actuator [13]

On the other hand, a fast long range actuator (FLORA) as shown in Fig. 2.8, can also be designed in such a manner that it utilizes an air-bearing slider and linear motors [15]. This FLORA can achieve a travel length up to 4 mm but at relatively low frequency of 20 Hz.

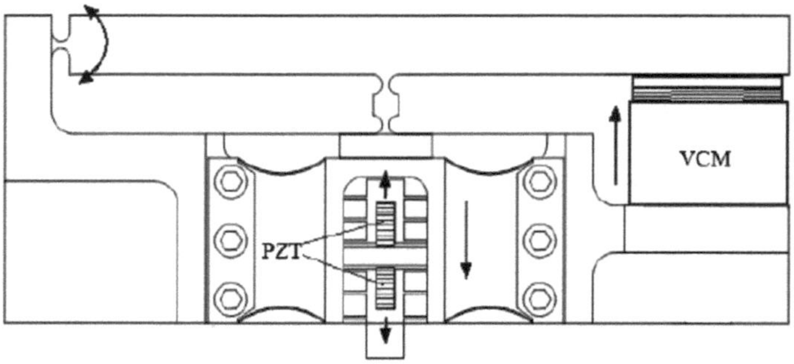


**Fig. 2.7** Hybrid PZT and voice coil FTS [14]

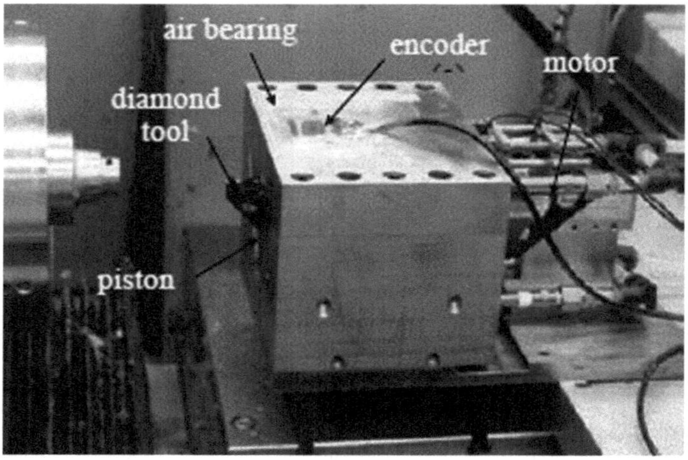


**Fig. 2.8** Photographic view of FLORA [14]

From the literature review, it can be concluded that long-stroke FTSs are usually actuated by piezoelectric (PZT) and voice coil actuators [10]. PZT FTSs are usually guided by flexure hinge structures which are more suitable for error compensation. However, piezoelectric FTSs often have a low resonance frequency because of its lever mechanism bringing hysteresis and tracking error. Voice coil FTSs may have longer travels but they have lower frequencies. Hence, the travel and the frequency are two separate performance parameters which cannot be optimized simultaneously for most cases [10].

### *2.1.2 Slow Slide Servo (SSS)*

Slow slide servo (SSS) diamond turning is engineered to address the travel limitation by the FTS system. SSS diamond turning has made its debut appearance in

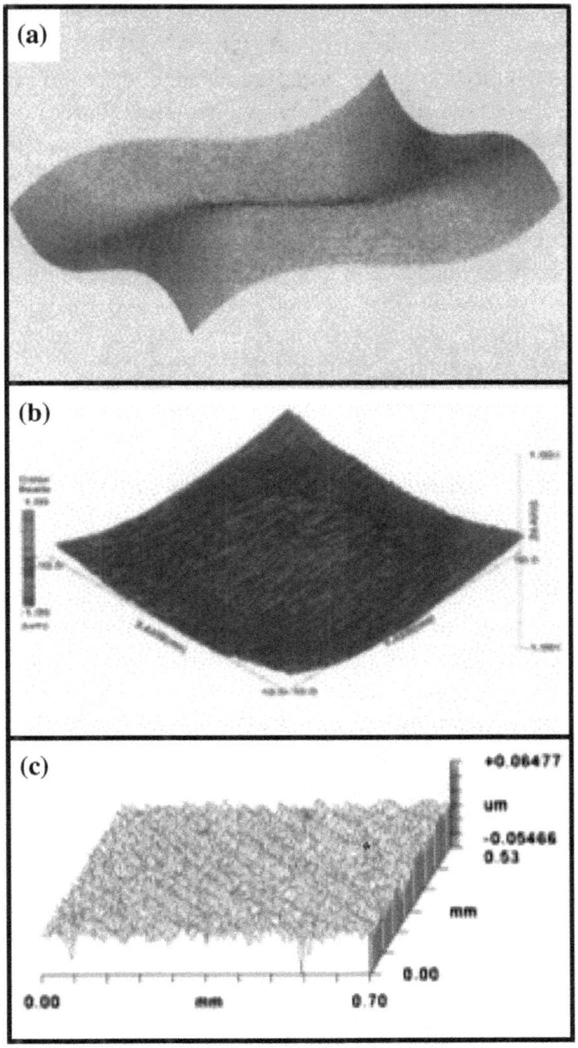


**Fig. 2.9** Cubic phase plate; **a** desired surface, **b** form accuracy of 0.263 μm, **c** RMS surface finish <5 nm [16]

2003 [16] and exhibited its distinguished performance to fabricate freeform surfaces, as illustrated in Fig. 2.9, exceeding 1 mm sag height with excellent surface quality and accuracy. This marks the tipping point for the growing interest of this novel ultraprecision machining technique to fabricate freeform optical surfaces with larger sag height. STS technology utilizes the existing diamond turning machine Z-slide for the tool motion by adopting linear motor to replace ball screws. This allows more flexibility in the motion of the slide without damaging the ball screw. It has advantages of fabricating parts with much larger deviation than the FTS. By exploiting its advantages, several works [17–19] have been carried out for the feasibility study of STS diamond turning to fabricate freeform optical surfaces with high accuracy and surface quality.

To the best of the author knowledge, there are very few works reported on SSS technology and this makes it vulnerable to unforeseen barriers in the machining of freeform surfaces. However, there are some issues in this SSS process which have not been addressed. Firstly, SSS process always faces a problem of having high inertial forces due to heavy-weight linear axes, which slows movements of machining axes. This makes it is not suitable for machining freeform surfaces with higher frequency and this results in lowering cutting speeds which would degrade the surface quality. Secondly, SSS is also plagued by thermal drift during the extremely long fabrication time, similar to those traditional fabrication methods such as grinding, polishing, or flycutting.

### *2.1.3 Other Multiple-Axis Ultraprecision Machining Techniques*

Figure 2.10 illustrates and classifies the complexity and scale of different machining processes for generating optical (micro-) structures with respect to their characteristic dimension [20]. Notwithstanding the fact that the multiple-axis ultraprecision machining techniques such as FTS, SSS, diamond milling, etc. have demonstrated their capabilities to fabricate hybrid freeform surfaces, they are still facing the difficulties to fabricate hybrid freeform surfaces with complex curvatures in a single setup. This is due to the increase of complexity associated with the loss

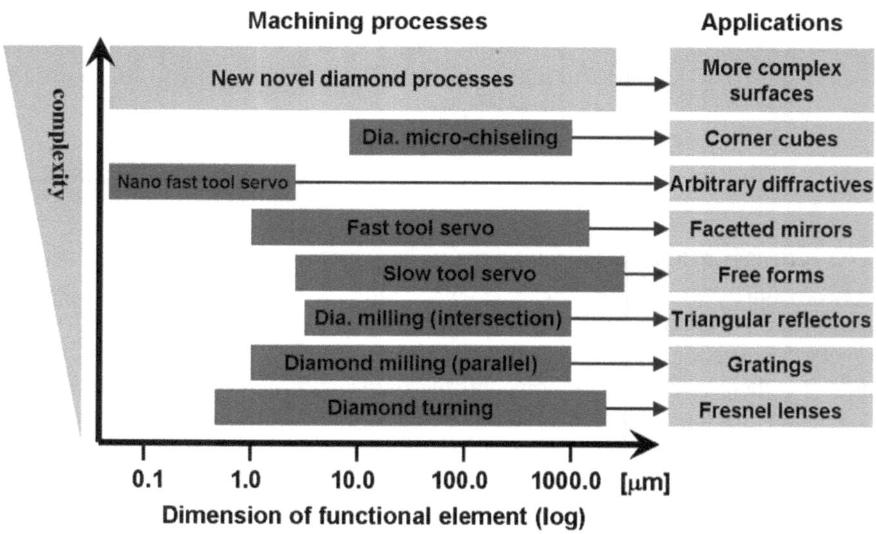


**Fig. 2.10** Complexity and dimension of optical (micro-) structures [20]

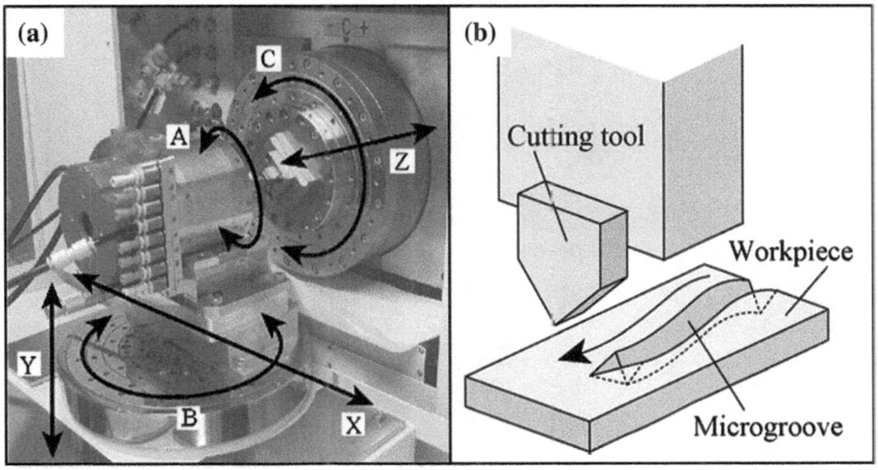


**Fig. 2.11** 6-axis grooving technique [23]; **a** configuration and **b** cutting a freeform groove

of symmetry of the surface [21]. Hence, it is necessary to increase the number of machining axes for moving a diamond tool to produce a freeform surface [22].

Recently, novel ultraprecision machining techniques have been developed to meet the increasing complexity for freeform surfaces in optical or mechanical applications. Multiple-axis microgrooving [23], as described in Fig. 2.11, is one of the novel techniques exhibiting the capability to fabricate flat-ended freeform microgrooves. Another novel technique is diamond micro-chiseling [24], as shown in Fig. 2.12, which has also demonstrated its capability to fabricate micro retro-reflectors and other micro-optical geometries. In the near future, more novel ultraprecision machining techniques would be begotten from the developments of ultraprecision machine systems offering a great opportunity to unlock the hidden potentials and challenges in the freeform optical surface.

One of the potential multiple-axis techniques is an interesting freeform engraving technique known as Guilloche engraving [25]. Guilloche engravings are widely used in luxury watches and jewelleries, and security features in passports, credit cards, etc. [26, 27]. Guilloche engraving utilizes a rose engine lathe [28], as shown in Fig. 2.13, to engrave a repetitive architectural patterns of intersecting or overlapping spirals [29, 30] in a spirograph manner. Spirograph [31] is a geometric drawing toy that produces mathematical roulette curves of the variety technically known as hypotrochoids and epitrochoids. Notwithstanding the fact that the Guilloche technique has the capability to produce freeform patterns, it is still a tedious and time consuming hand-operated process. Thus, an automated technique is required to replace this hand-operated process.

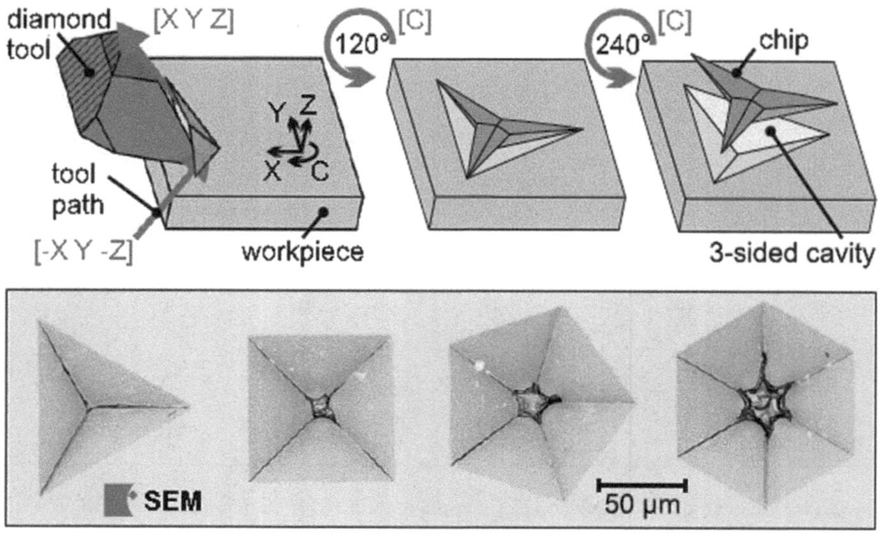


**Fig. 2.12** Diamond micro-chiseling technique [24]; (*top*) cutting mechanism and (*bottom*) micro-chiseled prismatic facets

**Fig. 2.13** **a** Rose engine lathe [28] and **b** Guilloche patterns [29]

## 2.2 State-of-Art CAD/CAM/CAE Technologies

### *2.2.1 CAD/CAM Technology for Surface Generation*

It is commonly to employ a computer-aided design (CAD) software (SolidWorks, etc.) to design the hybrid freeform surfaces for optical applications, instead using of traditional and tedious mathematical approaches. CAD software solutions provide not only designing solutions but also simulation analysis solutions with an aid of computer-aided engineering (CAE) technology [32]. These CAD models could be further utilized for post-processing into tool path by CAM software. However, there are common shortcomings for conventional CAM software solutions which make them unsuitable for generating tool path on freeform surfaces, which are of resolution ranges of 10 nm in the CAM systems and lack of post-processing system to support FTS/SSS processes [33]. This 10 nm resolution range not only is large as compared to ultraprecision applications (<1 nm), but also often causes large shape deviations and poor surface roughness on the fabricated freeform surface. Hence, customized CAM software is deemed necessary for providing suitable post-processor with adequate accuracy in generating tool path for FTS/SSS diamond turning.

Although there are available commercial software solutions such as DIFFSYS [34] and NanoCAM 2D/3D [35] for FTS/SSS diamond turning, they are still very costly and their methodologies for generating accurate surfaces with optimized process parameters are executed in the black box. Hence, there is still a room for improvement on the optimization of process parameters to achieve good surface quality and accuracy. Manufactures have been searching for solutions to sustain their competitive advantage in mass producing products at the shortest time to market and at a most economical cost. Hence, these drive the need for an alternative and economical option of generating accurate tool trajectory for FTS/SSS diamond turning and other multiple-axis machining processes.

Fortunately, the resolution ranges found in most commercial CAD software solutions are generally finer than those in CAM systems. This advantage could be further exploited for generating tool trajectory on freeform surfaces by employing the Visual Basic application programming interface (API). This marks the tipping point for growing attention on integration of API and CAD software for reverse engineering methodologies [36]. However, according to author's knowledge, there is no implementation of API approach to generate accurate spiral tool trajectories for the FTS and SSS process.

### 2.2.2 *Surface Accuracy and Error Compensation Approaches*

The new era of ultraprecision machining technologies for freeform optics requires the advancement of design and testing for innovative optical function and improved optical performance [37]. Although a mirror surface is necessary for a good optical performance of freeform surfaces, the surface accuracy is a dominant factor for the overall optical performance. Figure 2.14 clearly explains that a freeform surface with good accuracy would guide the lights to the designed paths accurately. Otherwise, the lights would be diverted away from the designated paths. Hence, much research work has been conducted in the area of surface generation with ultraprecise surface accuracy for freeform optics. Over the past decades, much

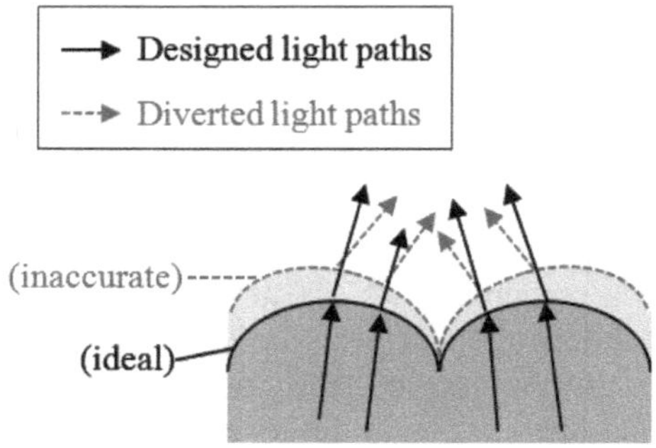


**Fig. 2.14** Effect of surface accuracy on the optical performance

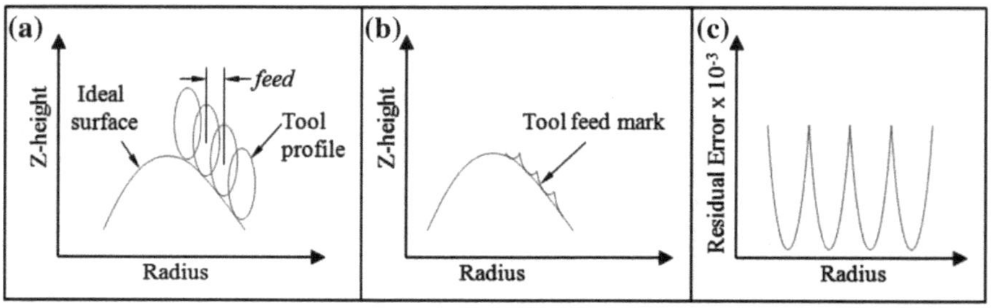


**Fig. 2.15** Cutting residual error of a machined freeform surface; **a** Tool profiles on ideal surface, **b** resulted tool feed marks on machined surface, and **c** resulted cutting residual errors due to the errors from feed marks

research has been conducted on the surface generation methods, machine dynamics, error analysis and methodologies for error compensation.

In the FTS/SSS processes, there are two different types of surface errors, namely cutting residual and cutting linearization errors. The cutting residual error, as described in Fig. 2.15, is the formation of tool marks on the surface along the feed direction [38–42]. Kong et al. [38] and Yu et al. [40] have studied and successfully developed their models for predicting this residual error in the FTS/SSS processes. They also conclude that the residual error dominates the errors by the tool nose radius and feedrate, and is analogous to surface roughness. Thus, a proper selection of feedrate and tool nose radius should be developed for machining accurate freeform surfaces.

In contrast, cutting linearization error, as illustrated in Fig. 2.16 is the Peak-to-Valley error ($PV_{err}$) between the ideal surface profile and the linear tool trajectory in the spiral cutting direction. This $PV_{err}$ depends mainly on the cutting distance between the two corresponding points in the cutting direction. It has been reported that the best profile accuracy results can be achieved by the spline interpolation method in the DiffSys software [43, 44]. However, the details for selecting cutting parameters in the spline interpolation method were not reported.

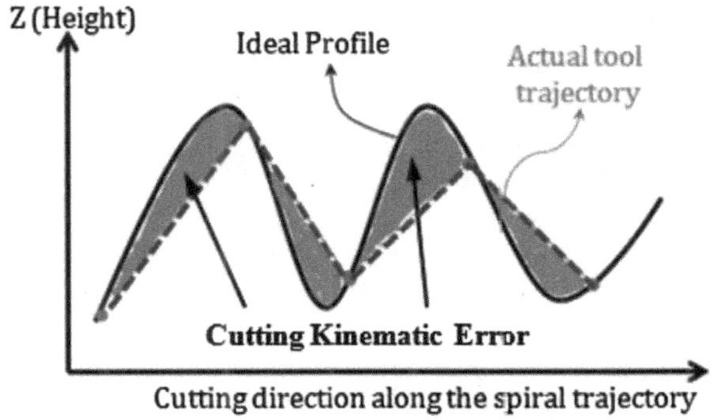


**Fig. 2.16** Cutting linearization error of a machined freeform surface

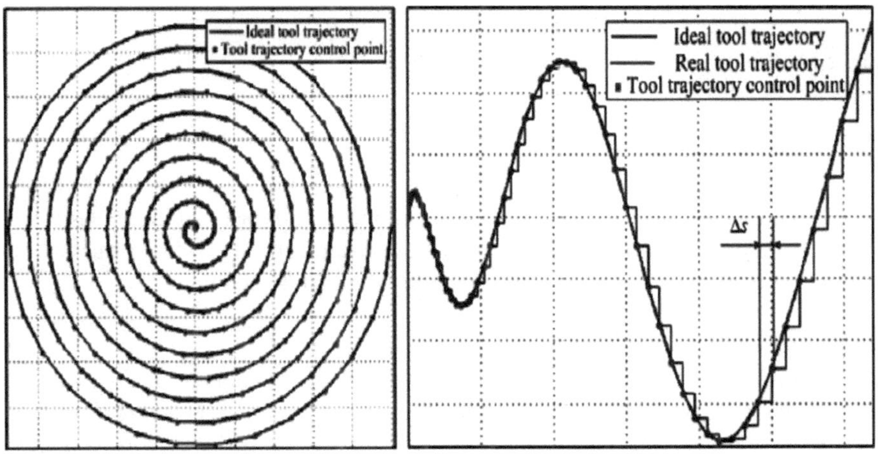


**Fig. 2.17** Tool trajectory by constant angle method [45]

Zhou et al. [45] have conducted a comparison studies on surface quality of machined surface based on constant-angle and constant arc-length methods, as illustrated in Figs. 2.17 and 2.18, respectively. When the constant-angle method is employed, the surface quality of outer regions is reported to be worse than that of central regions due to arc-lengths between the corresponding points on outer regions are sparser than those on central regions. Whereas the constant arc-length method shares the same number of cutting points as constant-angle method, the machined surface error is constant.

It has been also reported that the selection of critical incremental arc-lengths plays an important role in achieving accurate ultraprecision freeform surfaces [46]. Notwithstanding the fact that the constant arc-length method demonstrated its capability for better overall profile accuracy, only sinusoidal wave grid (SWG) profiles were served as case studies. It still remains unknown whether the constant arc-length method is also an ideal method for other profiles as well.

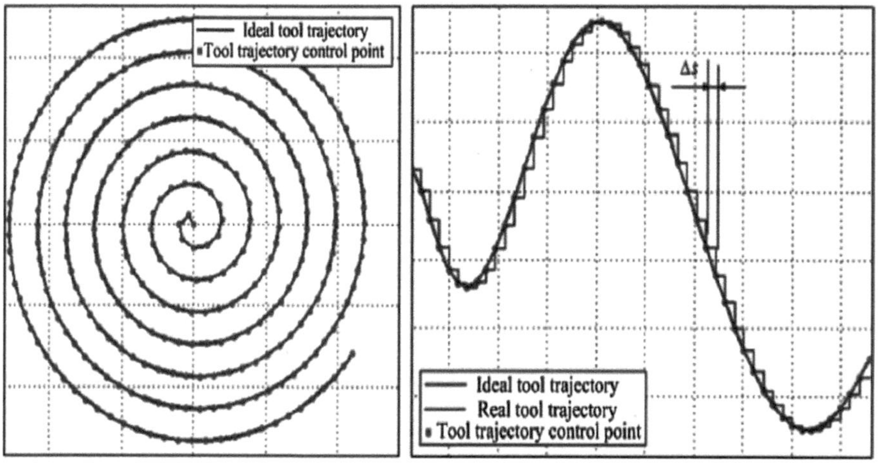


**Fig. 2.18** Tool trajectory by constant arc-length method [45]

## 2.3 Concluding Remarks

From the literature review, it can be concluded there are four main loopholes which are necessary to be addressed for the seamless manufacturing of freeform surfaces with good quality and accuracy. They are highlighted as below:

(i) Limitations of FTS and SSS processes

The limited travel length of fast tool servo (FTS) system is not suitable for machining freeform surfaces with larger sag heights. The travel length has to be increased by employing either piezoelectric and/or voice coil actuators. Although piezoelectric FTSs are more suitable for error compensation, they often have a low resonance frequency because of the lever mechanism. This lever mechanism also brings hysteresis and tracking error from the lever bending. On the other hand, voice coil FTSs may have longer travel than piezoelectric FTSs but they have lower bandwidth than other FTSs.

Although the slow slide servo (SSS) process may have much longer travel length than FTS process, the major limitation of SSS is having a low bandwidth system due to the heavily-weighted machine slides. This low bandwidth makes SSS process not suitable for machining freeform surfaces with higher frequency asymmetries. It also limits the cutting speed to low spindle speeds which would bring the possibility of degrading the surface finish. Furthermore, the low spindle speed may lead to thermal drifts as the fabrication time of the freeform surface increases.

Both of the travel and the bandwidth are two separate performance parameters which cannot be optimized simultaneously in most cases. Hence, a study for optimization of FTS and SSS process is necessary to obtain both optimal travel and bandwidth for the machining of hybrid freeform surfaces.

(ii) Machining barriers in the machining of complex freeform surfaces

Although FTS and SSS diamond turning can produce complex three dimensional structures, higher degrees of freeform optical surfaces such as Fresnel lens arrays are yet to be achieved in a single setup. This is due to the increase of complexity associated with a loss of symmetry of the surface. Hence, it is necessary to increase the number of machining axes (degrees of freedom) for moving a diamond tool overcoming the loss of symmetries to produce a freeform surface. Few novel ultra-precision diamond machining techniques have been begotten from the developments of ultraprecision machines to fulfil a great deal of demands for generating highly complex freeform optical surfaces.

However, these machining techniques are suitable for only a handful of freeform surfaces and may require several setups which would lead to accumulation of setting errors resulting inaccurate ultraprecision freeform surfaces. Thus, a novel multiple-axis ultraprecision machining technique is necessary to minimize the number of setups for producing accurate ultraprecision hybrid freeform surfaces.

(iii) Cutting linearization errors

Cutting linearization error is the profile error ($PV_{err}$) between the ideal surface profile and the linear tool trajectory in the spiral cutting direction. This inherited error should not be ignored as it would sequentially accumulate those errors from the machine dynamics, etc., resulting larger errors or poorer profile accuracy. Cutting linearization error usually depends on the machinist's skills and experience, and it often analyzes during post-machining stages. Hence, it is not reliable to depend on highly skill labour and not cost effective to analyze in post-machining stages. Therefore, it is necessary to develop a model to analyze the cutting linearization error for generating accurate ultraprecision freeform surfaces in the FTS and SSS processes.

(iv) Comprehensive CAD/CAM system for machining of hybrid freeform surfaces

Most of commercial CAM software solutions (SolidCAM, Unigraphics, etc.) are presently unable to support fast tool/slow slide servo (FTS/SSS) diamond turning of freeform surface due to the difference in the coordinating systems. Hence, a special CAM post-processor is required to generate the spiral tool trajectory in the FTS/SSS diamond turning. However, these special CAM software solutions are very expensive. Manufactures have been searching for solutions to sustain their competitive advantage in mass producing products at the shortest time to market and at a most economical cost. Hence, these drive the needs for an alternative and economical option of generating accurate tool trajectory for FTS/SSS diamond turning of freeform surfaces.

## References

1. Davis GE, Roblee JW, Hedges AR. Comparison of freeform manufacturing techniques in the production of monolithic lens arrays. Proc SPIE 2009;7426 742605-1.
2. http://www.iiviinfrared.com.

3. Huang C, Li L, Yi AY. Design and fabrication of a micro Alvarez lens array with a variable focal length (Technical paper). Microsystems Technology 2009;15:559–63.
4. Li L, Yi AY. Microfabrication on a curved surface using 3D microlens array projection. J Micromechanical Microeng 2009;19:105010.
5. Li L, Yi AY. Design and fabrication of a freeform prism array for 3D microscopy. J Opt Soc Am A 2010;27(12).
6. Li L, Yi AY. Design and fabrication of a freeform microlens array for a compact large-field-of-view compound-eye camera. Appl Opt 2012;51(12).
7. Patterson SR, Magrab EB. Design and testing of a fast tool servo for diamond turning. Precis Eng. 1985;7(3):123–8.
8. Meinel AB, Meinel MP, Stacy JE, Saito TT, Patterson SR. Wavefront correctors by diamond turning. Appl Opt 1989;25.
9. Luttrell DE. Machining non-axisymmetric optics. In ASPE proceedings, annual meetings, 1990.
10. Neo WK, Kumar AS, Rahman M. A novel method for layered tool path generation in the fast tool servo diamond turning of non-circular microstructural surfaces. Proc Inst Mech Eng Part B J Eng Manuf 2013;227(2):210–19.
11. Ludwick SJ, Chargin DA, Calzaretta JA, Trumper DL. Design of a rotary fast tool servo for ophthalmic lens fabrication. Precis Eng. 1999;23:253–9.
12. Kim HS, Lee KI, Lee KM, Song BS. Fabrication of free-form surfaces using a long-stroke fast tool servo and corrective figuring with on-machine measurement. Int J Mach Tools Manuf. 2009;49:991–7.
13. Rakuff S, Cuttino JF. Design and testing of a long-range, precision fast tool servo system for diamond turning. Precis Eng. 2009;33:18–25.
14. Liu Q, Zhou X, Wang L. A new hybrid macro- and micro-range fast tool servo. In: 2010 International conference on mechanic automation and control engineering (MACE).
15. Buescher N, Dow TA, Sohn A, Norlund B, Roblee J. Live-axis turning. In: ASPE proceedings, annual meeting, 2004.
16. Tohme YE, Lowe JA. Machining of freeform optical surfaces by slow slide servo method. In: Proceedings of the ASPE, annual meetings, 2003.
17. Li L, Yi AY, Huang C, Grewell DA, Benatar A, Chen Y. Fabrication of diffractive optics by use of slow tool servo diamond turning process. Opt Eng. 2006;45(11):113401.
18. Li L, Yi AY. Development of a 3D artificial compound eye. Opt Express 2010;18(17).
19. Li LK, Yi AY. Design and fabrication of a freeform microlens array for uniform beam shaping (Technical paper). Microsyst Technol 2011;17:1713–20.
20. Flucke C, Osmer J, Lünemann B, Riemer O, Brinksmeier E. Scaling in machining of optics with reflective to diffractive function. In: Proceedings of the 9th international Euspen conference. vol. 1, 2009, p. 17–20.
21. Riemer O. Advances in ultra precision manufacturing. In: International symposium of the Japan Society for Precision Engineering 2011 (ISUPEN 2011).
22. Neo DWK, Kumar AS, Rahman M. Automated Guilloche machining technique for the fabrication of polygonal Fresnel lens array. Precis Eng. 2015;41:55–62.
23. Moriya T, Nakamoto K, Ishida T, Takeuchil Y. Creation of V-shaped microgrooves with flat-ends by 6-axis control ultraprecision machining. CIRP Ann Manuf Technol 2010;59: 61–66.
24. Flucke C, Gläbe R, Brinksmeier E. Diamond micro chiseling of large-scale retroreflective arrays. Precis Eng. 2012;36:650–7.
25. Guilloche, Wikipedia. http://en.wikipedia.org/wiki/Guilloche.
26. van Renesse RL. Paper based document security-a review, European Conference on Security and Detection (ECOS 97), IEE, London, United Kingdom, 28–30 April 1997, Conference Publication No. 437, doi:10.1049/cp:19970425.
27. Trueb LF. Gold in watchmaking. Gold Bull. 2000;33(1):11–24. doi:10.1007/BF03215478.
28. Rose engine lathe, Wikipedia. http://en.wikipedia.org/wiki/Rose_engine_lathe.

29. Warfield B. Details: essential for successful CNC design, Blog, CNC Projects, Cool, Jan 7 2014. http://blog.cnccookbook.com/2014/01/07/details-essential-for-successful-cnc-design/.
30. Master Guillocheur. https://www.youtube.com/watch?v=88s98-SHNUU.
31. Spirograph, Wikipedia. http://en.wikipedia.org/wiki/Spirograph.
32. Garcia KJ. Non-rational and rational parametric descriptions of the geometric propagation of light in an optical system. Ph.D. Dissertation, The University of Arizona, 1999.
33. Steinkopf R, Dick L, Kopf T, Gebhardt A, Risse S, Eberhardt R. Data handling and representation of freeform surfaces. Proc SPIE 8169:81690X-2.
34. DIFFSYS. www.diffsys.com.
35. NanoCAM 2D/3D. http://www.nanotechsys.com/accessories/nanotech-250upl-additional-accessories/.
36. Gattamelata D, Pezzuti E, Valentini PP. Using application programming interface to integrate reverse engineering methodologies into SolidWorks. In: XVIII Congreso Internacional de Ingeniería Gráfica de INGEGRAF, 2006.
37. Thompson KP, Rolland JP. Freeform optical surfaces: a revolution in imaging optical design. Opt Photonics News. 2012;23(6):30–5.
38. Kong LB, Cheung CF. Modeling and characterization of surface generation in fast tool servo machining of microlens arrays. Comput Ind Eng. 2012;63(4):957–70.
39. Kwok TC, Cheung CF, Kong LB, To S, Lee WB. Analysis of surface generation in ultra-precision machining with a fast tool servo. Proc IMechE, Part B J Eng Manuf. 2010;224 (9):1351–67.
40. Yu DP, Hong GS, Wong YS. Profile error compensation in fast tool servo diamond turning of micro-structured surfaces. Int J Mach Tools Manuf. 2012;52(1):13–23.
41. Yin ZQ, Dai YF, Li SY, Guan CL, Tie GP. Fabrication of off-axis aspheric surfaces using a slow tool servo. Int J Mach Tools Manuf. 2011;51(5):404–10.
42. Yu DP, Gan SW, Wong YS, Hong GS, Rahman M, Yao J. Optimized tool path generation for fast tool servo diamond turning of micro-structured surfaces. Int J Adv Manuf Technol. 2012;63(9):85–99.
43. Scheiding S, Yi AY, Gebhardt A, Li L, Risse S, Eberhardt R, Tünnermann A. Freeform manufacturing of a micro-optical lens array on a steep curved substrate by use of a voice coil fast tool servo. Opt Express. 2011;19(24):23938–51.
44. Dick L, Risse S, Tünnermann A. Injection molded high precision freeform optics for high volume applications. Adv Opt Technol. 2012;1:39–50.
45. Zhou M, Zhang HJ, Chen SJ. Study on diamond cutting of nonrationally symmetric microstructured surfaces with fast tool servo. Mater Manuf Proc. 2010;25(6):488–94.
46. Liu K, Wu H, Liu P, Shaw KC. Ultra-precision machining of aluminium alloy surfaces for optical application. Int J Nanomanuf. 2011;7(2):116–25.

# Chapter 3
# Initial Development of CAD/CAM Technologies

This chapter presents an initial development of CAD/CAM technologies which implement Visual Basic application programming interface (API) into the SolidWorks platform to generate accurate toolpaths for diamond turning of hybrid freeform surfaces.

CAD software (SolidWorks, etc.) solution is widely preferred over traditional and tedious mathematical approaches for not only ease designing and modelling of hybrid freeform optical surfaces, but also simulating the analytical solutions for both imaging and non-imaging systems [1]. On the other hand, CAM software would post-process these CAD models into a tool path for machining. However, the conventional CAM software solutions are not suitable for generating ultraprecise tool path (decimal resolutions finer than 10 nm) and their post-processing system are still unable to support FTS/SSS processes [2]. Tool path with fewer decimal places causes not only poor surface roughness but also inaccurate surface profiles. Thus, a customized CAM software or post-processor generator is a necessity tool to generate accurate tool path of FTS/SSS diamond turning.

Although there are available commercial software solutions such as DIFFSYS [3] and NanoCAM 2D/3D [4] for FTS/SSS diamond turning, they are very costly. As there is a great intense competition in the market, manufacturers have been searching for solutions to sustain their competitive advantage in mass producing products at the shortest time to market and at a more economical cost. Hence, these drives the needs for an alternative and economical option of generating accurate tool trajectories for FTS/SSS diamond turning of freeform surfaces.

Fortunately, the decimal resolution ranges in CAD software solutions are finer than those in CAM systems and could be further exploited for generating a tool path on freeform surfaces with an aid of Visual Basic application programming interface (API). Furthermore, most CAD software solutions have their own built-in API environment allowing integrating predefined native geometrical entities and oper-

Some content of this chapter has been reproduced with permission from [12].



D.W.K. Neo, *Ultraprecision Machining of Hybrid Freeform Surfaces Using Multiple-Axis Diamond Turning*, Springer Theses,
DOI 10.1007/978-981-10-4083-2_3

ations with any user-defined computational algorithms. This marks the tipping point of growing attention on the integration of API into the CAD software for reverse engineering methodologies [5]. Thus, it also offers an attractive solution for generating accurate tool path on freeform surfaces other than CAM systems. In this proposed methodology, the spiral tool trajectory can be generated directly from the CAD models of freeform surface without utilizing any expensive CAM software.

## 3.1 CAD/CAM for Multiple-Axis Ultraprecision Machining Processes

### 3.1.1 Non-uniform Rational B-Spline Freeform Surfaces

Freeform surfaces are generated either based on the derived equations or CAD. Traditionally, the faceted surfaces with sharp edges and free-forms are defined by the means of section-functional methods. This traditional definition not only unable to be analytically closed, but is also difficult to solve numerically [6]. Non-uniform rational B-splines (NURBS) allow a much more elegant and flexible definition of surfaces, providing an implicit method of function switching capable of both realizing sharp edges and continuous transitions. Thus, NURBS are a class of splines having ideal characteristics for the specification of machine tool paths. A NURBS based surface $S(u, v)$ is described in Eq. (3.1) and its rational basis functions, as in Eq. 3.2, are weighted $p$th and $q$th degree B-Spline basis functions $N_{i,p}$ with the parameters $u$ and $v$.

$$S(u,v) = \sum_{i=0}^{n} \sum_{j=0}^{m} R_{i,j}(u,v)\, P_{i.j} \tag{3.1}$$

$$R_{i,j}(u,v) = \frac{N_{i,p}(u) N_{j,q}(v) w_{i,j}}{\sum_{k=0}^{n} \sum_{l=0}^{m} N_{k,p}(u) N_{l,q}(v) w_{k,l}} \tag{3.2}$$

The knot vector $N_{i,\ p}$ determines the activation control points on the NURBS surface while the order of a NURBS surface determines a number of control points influencing any given point on the surface. It is common knowledge that the conventional machine tool system cannot demystify directly from this description of a freeform surface. Therefore, it is deemed necessary to pre-process the surface data into tool trajectory with point clouds. In the fast tool/slow slide servo diamond turning, the point clouds are given in a spiral point $P_{i,\ j}$ in the XY-plane. For the tool path generation the z-component and the normal vector for every point $P_{i,\ j}$ have to be determined for the surface.

However, it would be tedious and time consuming process to calculate the intersection point between the line from Point $P$ and the surface $S(u,\ v)$ and the normal vector of the surface. Thanks to the latest developments in the CAD system, the intersection points and the normal vectors could be extracted directly from the NURBS surface data without tedious computations. The details for extracting these intersection points and normal vectors in the CAD system are discussed in the Sect. 3.2.

### *3.1.2 CAD/CAM Interpolator for FTS/SSS Diamond Turning*

Fast tool/slow slide servo (FTS/SSS) diamond turning is often employed for machining a freeform surface with a high degree of complexity. FTS diamond turning integrates a high bandwidth servo unit in an additional W-axis (or super-imposed Z-axis) with the existing three axes (X, Z and C-axis) in ultraprecision turning machine. SSS diamond turning is similar to the FTS technique except that it employs an existing Z-axis of the machine to oscillate the tool, but at lower bandwidth due to its inertia.

In the process of surface generation, the tool trajectory evolves as an Archimedes spiral which usually proceeds from outer radius to the center of workpiece. Since the spiral tool trajectory begins at the outer radius $r$, the radial position of the tool undoubtedly decreases as tool feeds towards the center. According to Weisstein [7], the radial position $\rho$ of the tool is given by:

$$\rho = r - f_r\theta \tag{3.3}$$

$$r = f_r\theta_t = f_r \times 2\pi N_t \tag{3.4}$$

$$N_t = r/(2\pi f_r) \tag{3.5}$$

where $f_r$ is the radial feed per radian, $\theta$ is the rotational angle of the machine spindle, $\theta_t$ and $N_t$ are the total angular and the total number of spiral rotations to reach the center from the outer radius, respectively. Hence, the spiral point $P$ of the tool trajectory in Fig. 3.1 is represented by polar coordinates as:

$$x = \rho \sin\theta = (r - f_r\theta)\sin\theta \tag{3.6}$$

$$y = \rho \cos\theta = (r - f_r\theta)\cos\theta \tag{3.7}$$

$$z \text{ or } W = f(\rho, \theta) = f(x, y) \tag{3.8}$$

where $P$ contains the coordinates of $x$, $y$ *and* z or $W$, and both $z$ and $W$ axes are the normal heights for the SSS and FTS processes, respectively, from the XY-plane to the intersection point on the surface. This spiral tool trajectory algorithm shall be

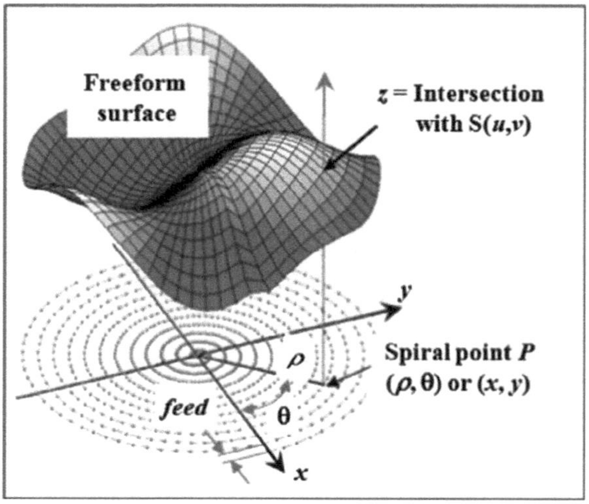


**Fig. 3.1** Archimedes spiral tool trajectory in FTS/SSS diamond turning. Figure reproduced from [12]

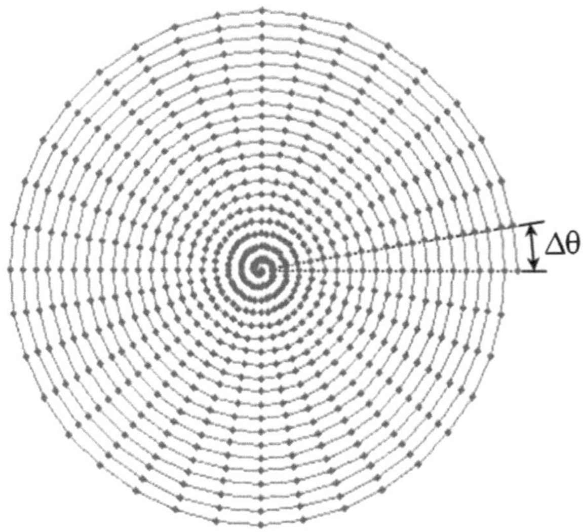


**Fig. 3.2** Constant-angle method of controlling tool trajectory

implemented into the API methodology for computation calculation and its details are also discussed in the later sections.

A common method for generating tool trajectory control points is constant-angle method, as explained in Fig. 3.2, and the value of constant angle positioning per revolution is given as:

$$\Delta\theta = 360/\mathrm{N}_p \tag{3.7}$$

where $N_p$ is the number of trajectory points per revolution.

There is an important factor to be considered for a good tool trajectory control, which is notwithstanding of constant angular positioning the arc length $\Delta S$ between two corresponding points becomes sparser as the radial position gets longer. If $N_p$ is too few, the surface quality of the outer portion would be lower than that of the inner portion. Hence, $N_p$ has to increase to reduce $\Delta S$ spacing giving a better surface quality at outer portion.

However, the number of control points is contrastively restricted by close-looped bandwidth of FTS/SSS system [8]. Large $N_p$ also reduces the rotational spindle speed and may face difficulty in cutting ductile metals which required high cutting speed. Hence, it is necessary to select the appropriate sampling number for good tool trajectory control and surface quality.

## 3.2 API Methodology for CAD/CAM Software Development

SolidWorks is a one of the widely used commercial software solutions bundled with several packages including application programming interface (API) for designing freeform surfaces. In this study, a SolidWorks-API package has been developed to generate spiral tool trajectories for the FTS/SSS diamond turning process. This developed package allows the users to input the necessary cutting parameters, namely spindle speed, feedrate, depth of cuts, tool nose radius, etc., for post-processing a freeform surface into tool trajectory points which are executed into NC codes.

This post-processing process, which is known as tool-interpolator, works in the developed API environment where the standard utilities, functions and user-defined algorithms are executed. The developed API allows the implementation of any user-defined algorithm to perform accurate computation by integrating predefined native geometrical entities and operations with a powerful Mathematical Utility, as explained by Gattamelate et al. [5]. This mathematical utility feature allows easily performing basic and advanced point, vector and matrix operations.

Figure 3.3 describes three groups of objects which can be manipulated with API in SolidWorks, namely native geometrical, native mathematical and user defined entities. The native geometrical entities include the sketch entities (point, line,

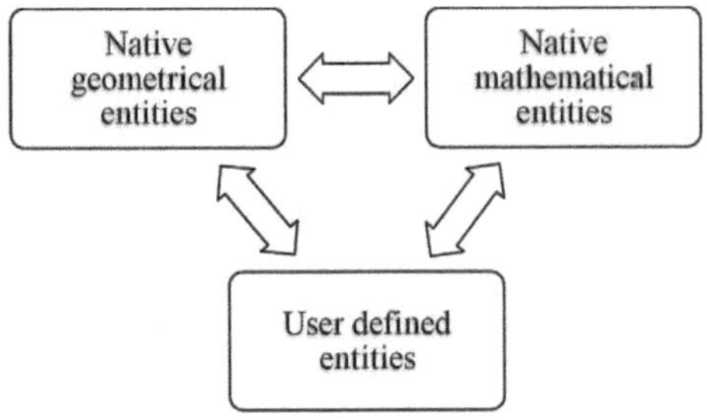


**Fig. 3.3** SolidWorks API entity scheme. Figure reproduced from [12]

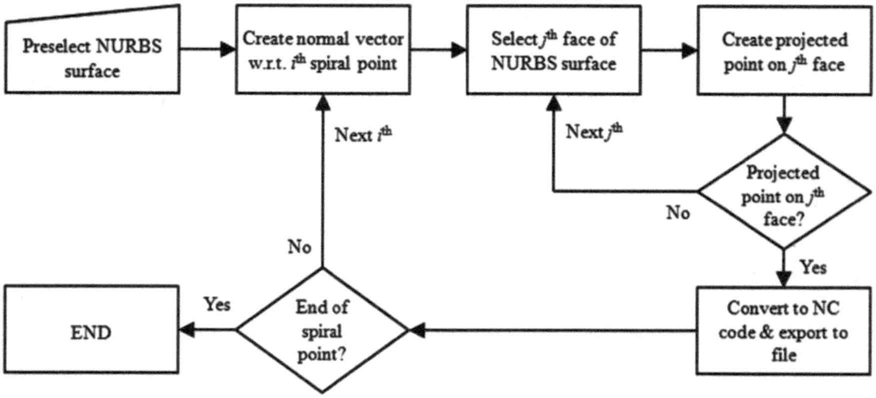


**Fig. 3.4** Process flow for computing z-value of intersection point on NURBS surface

circle, spline, etc.) and their constraints, the features (extrusion, revolution, loft, etc.) and the assembly management (mating, inserting, moving, etc.). The mathematical native entities such as points, vectors and transformations are manipulated for projecting from model space to sketch space and vice versa, performing basic operation on vectors, etc. It is also recommended that a direct accessing into internal database of entities to execute commands and interlace model entities with math and user-defined entities [5]. This would give an optimized handling for extremely large amount of data points.

In this study, the process flow, as illustrated in Fig. 3.4, has been developed and implemented to determine the $z$-values of the spiral intersection points on the NURBS surface.

Firstly, the mathematical utility performs point operations to create points and vectors in the CAD database with respect to the spiral points $P$ by using the existing built-in API functions, namely '*CreatePoint*' and '*CreateVector*', as demonstrated in Fig. 3.5.

However, a NURBS surface in a CAD model is usually associated with the multiple faces, as illustrated in Fig. 3.6, which are necessary to be geometrically defined by using built-in geometrical entity API function '*GetFaces*'. This '*GetFaces*' function identifies those faces associated to a NURBS surface. These identified faces are necessary to overcome the loss of data during the point-intersection stage.

The evaluated normal vector intersects onto a selected face of NURBS surface by employing a geometrical entity function '*GetProjectedPointOn*'. However, these faces are randomly selected and may not have a resulted projected point on a randomly selected face. Hence, it is necessary to perform iteration to select the next 'face' until a resulted projected point is obtained, as shown in Fig. 3.7. This iteration process is to prevent the loss of data during the generation of spiral tool trajectory on the freeform surface.

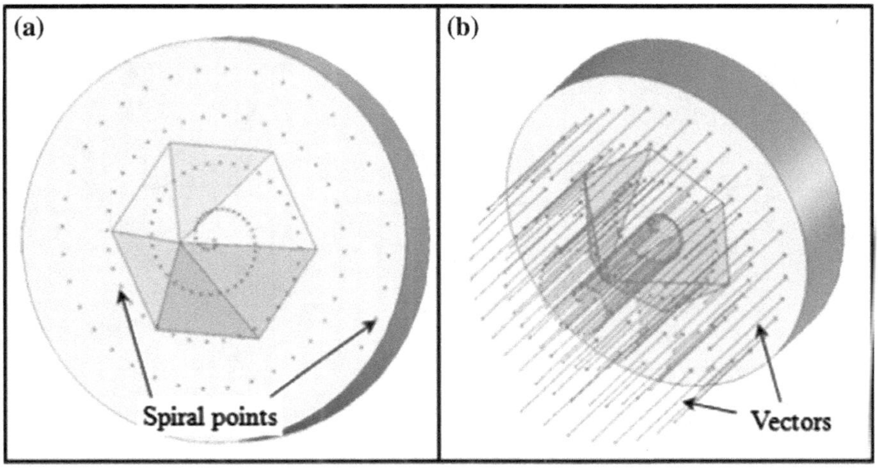


**Fig. 3.5** Mathematical utility in API; **a** CreatePoint and **b** CreateVector functions

**Fig. 3.6** Multiple faces in a desired surface were identified by using GetFaces function

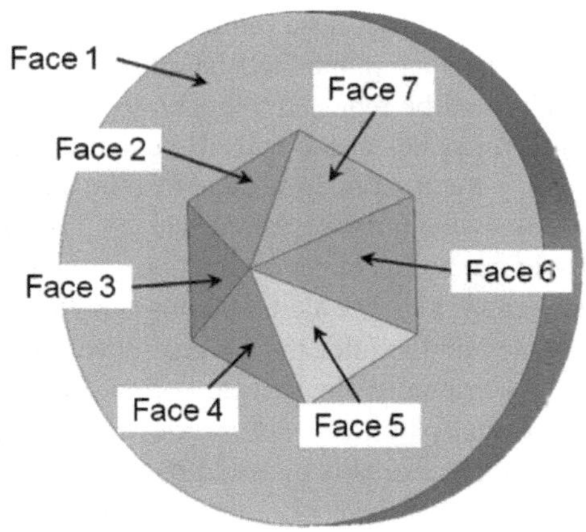


**Fig. 3.7** Defining an intersection point on the NURBS surface using 'GetProjectedPointOn' function

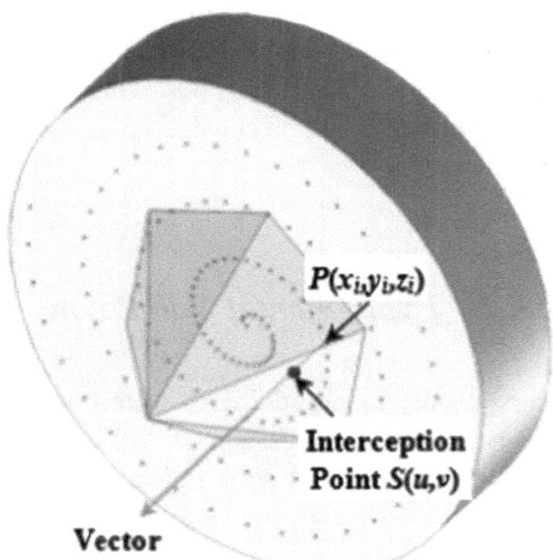

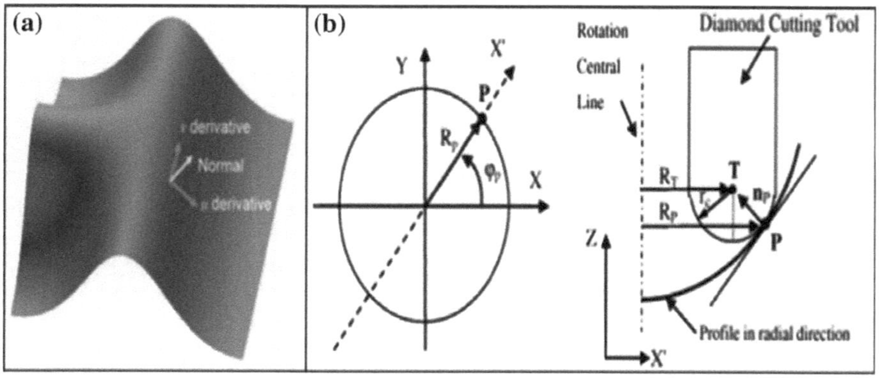


**Fig. 3.8** Traditional computations of surface normal; **a** definition [10], and **b** tool nose radius compensation [11]

Finally, the resulted projected point is also known as one of the point clouds on NURBS surface, which can be further processed into a spiral tool trajectory control point with a tool nose radius compensation methodology. Then, this compensated tool trajectory point shall be exported to a NC file for machining processes. This cycle shall be repeated to generate every spiral point cloud on the NURBS surface until it reaches at the end of the spiral trajectory.

It is a common knowledge that the effect of tool nose radius on machined surface accuracy cannot be ignored. Hence, it is deemed necessary to compensate for the tool nose radius effect on the generation of tool trajectory control points. If the tool trajectory is generated without the tool nose radius compensation, the surface profile would suffer the adverse effect of overcutting problem [9].

Traditionally, the surface normal $N(u, v)$ of a NURBS surface has to be defined numerically by means of a series of calculations [10, 11], as described in Fig. 3.8. A surface normal is essential to compute the tool nose radius compensation for the generation of the tool trajectory to prevent the overcuts.

Fortunately, this tedious computation could be eliminated with an aid of a modelling utility "*offset surface*". This utility offsets a surface feature by amount of offset value equivalent to tool nose radius $T_R$ in the normal direction of the NURBS surface, as shown in Fig. 3.9. Therefore, the projected points $S(u, v)_{offset}$ on the offset surface could be considered as the tool trajectory points with tool nose radius compensation.

### 3.2.1 *Experimental Validation*

The proposed methodologies were implemented into the SolidWorks CAD system by using its Visual Basic 6.0 API environment to verify their capabilities to generate accurate spiral tool trajectories. In the experiments, the slow slide servo

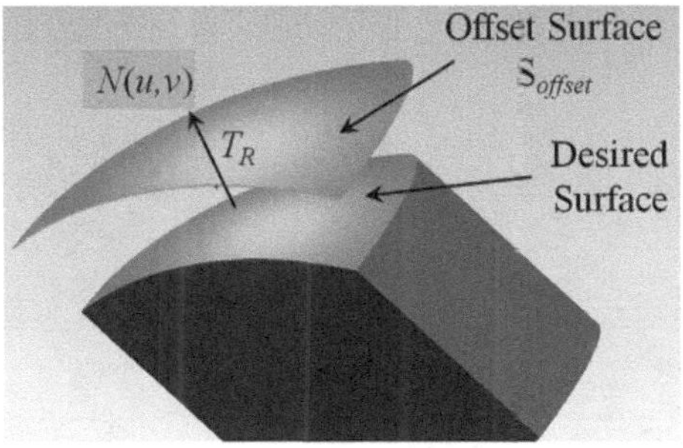


**Fig. 3.9** Tool nose radius compensation can be simplified by offsetting a NURBS surface with modelling utility, "offset surface"

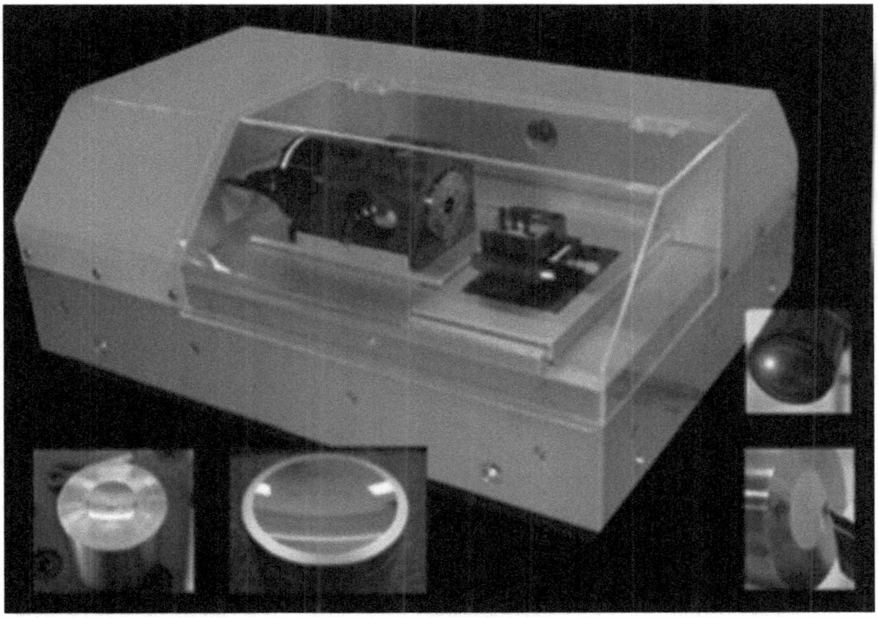

**Fig. 3.10** Photographic image for miniature ultraprecision lathe UPL-420

(SSS) diamond turning technique is being employed with Mikrotools ultraprecision lathe UPL-420, as shown in Fig. 3.10, to fabricate a multiple-compound eye surface, as illustrated in Fig. 3.11. The parameters, as described in Table 3.1, are also selected for the cutting conditions in the experiments, which are also been employed as the inputs for the surface generation of spiral tool trajectory in the developed SolidWorks-API user interface.

Figure 3.12 displays a successful surface generation of spiral tool trajectories for the FTS/SSS process based on the cutting parameters as in Table 3.1, which were

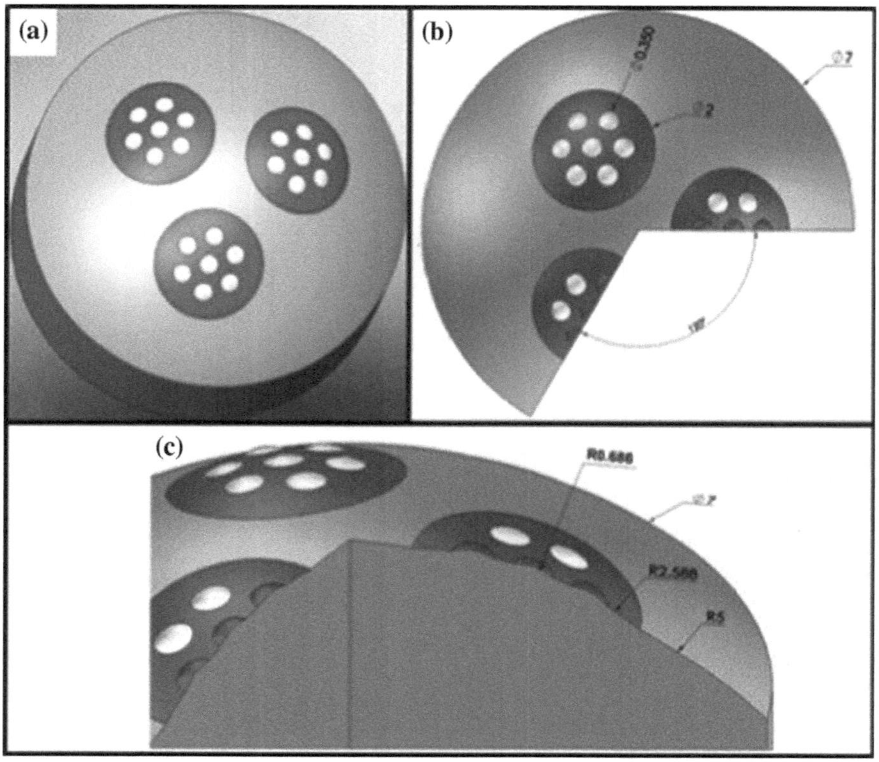


**Fig. 3.11** CAD model of multiple compound eye Design-A, **a** 3D view, **b** top view, and **c** slanted half-section view

**Table 3.1** Cutting conditions for fabrication multiple compound eye design-A

| | |
|---|---|
| Feedrate | 0.005 mm/rev |
| Spindle speed | 25 rev/min |
| $\Delta\theta$ ($N_p$) | 1.0° (360 points per revolution) |
| Depth of cut | 1–5 μm (finish); 10 μm (rough) |
| Workpiece | Brass<br>Radius 3.5 mm |
| Tool | Mono-crystal diamond<br>$T_R$ = 0.2 mm |

mapped onto the surface of the CAD model with the implementation of the proposed methodologies in the developed SolidWorks-API user-interface, as illustrated in Fig. 3.13. The details of the applications and the outputs by the developed SolidWorks-API CAD/CAM system are presented and discussed with case studies in the Chap. 7.

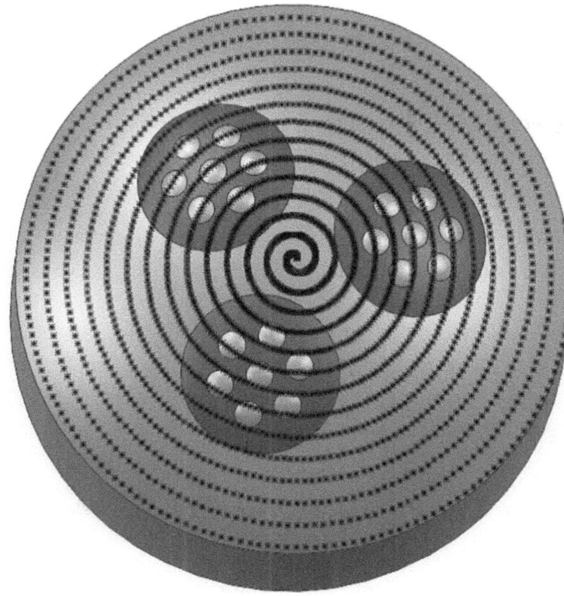

**Fig. 3.12** A successful generation of spiral tool trajectories for FTS/SSS diamond turning process, which are mapped onto the surface of the CAD model. (*Note* The number of spiral trajectory points had been reduced for a better clarity)

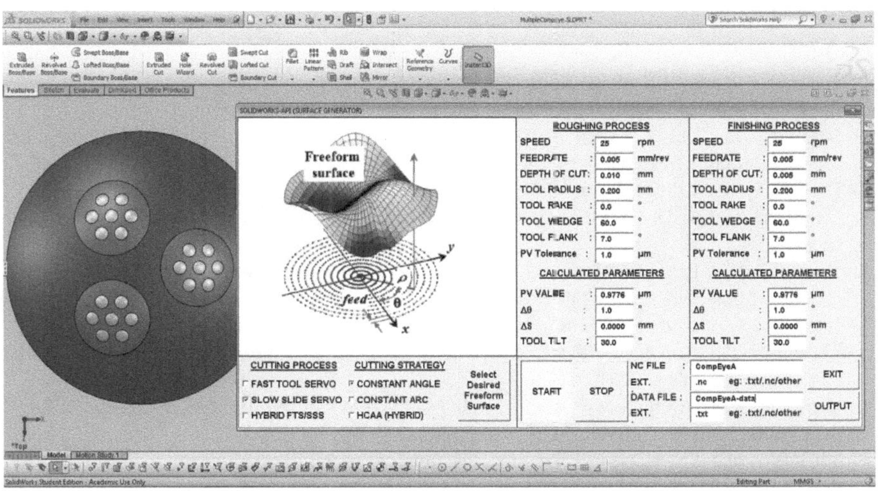


**Fig. 3.13** A screenshot image for the user-interface in the developed SolidWorks-API CAD/CAM system

## 3.3 Concluding Remarks

In the present study, the initial development of CAD/CAM tool interpolator for fast tool/slow slide servo (FTS/SSS) diamond turning of hybrid freeform surface shows that expensive customized CAM system is no longer necessity tool for the tool path generation of freeform surfaces. From the results, it concludes that:

i. The integration of application programming interface (API) into the SolidWorks environment has been validated with the successful machining of hybrid freeform surface (multiple-compound eye).
ii. The proposed methodologies only use the given properties and information provided by the surfaces in CAD model directly. This eases the needs for tedious and time consuming process to calculate numerically for both of the intersection point between the normal vector from Point $P$ and the surface $S(u, v)$.

However, there are few areas which are required to be addressed for improvements. They are as follows:

i. The cutting speed in the SSS process was too slow which degraded the surface finishing. Hence, a hybrid FTS/SSS process has been developed to address this cutting speed issue and the details are presented in the next chapter.

## References

1. Garcia KJ. 'Non-rational and rational parametric descriptions of the geometric propagation of light in an optical system. Ph.D. Dissertation, The University of Arizona, 1999.
2. Steinkopf R, Dick L, Kopf T, Gebhardt A, Risse S, Eberhardt R. Data handling and representation of freeform surfaces. In: Proceedings of SPIE Vol. 8169 81690X-2.
3. DIFFSYS, www.diffsys.com.
4. NanoCAM 2D/3D, http://www.nanotechsys.com/accessories/nanotech-250upl-additional-accessories/.
5. Gattamelata D, Pezzuti E, Valentini PP. Using application programming interface to integrate reverse engineering methodologies into solidworks. In: XVIII Congreso Internacional de Ingeniería Gráfica de INGEGRAF, 2006.
6. Thompson KP, Rolland JP. Freeform optical surfaces: a revolution in imaging optical design. Opt Photonics News. 2012;23(6):30–5.
7. Weisstein EW. Archimedes' spiral, From MathWorld—A Wolfram Web Resource, http://mathworld.wolfram.com/ArchimedesSpiral.html.
8. Yu DP, Wong YS, Hong GS. Optimal selection of machining parameters for fast tool servo diamond turning. Int J Adv Manuf Technol. 2011;57:85–99.
9. Hsu WY, Chen FZ, Kuo CH, Cheng YC, Lin WL. Fabrication of micro aspherical lens array using fast tool servo with tool path compensation. Annual Meetings: ASPE Proceedings; 2008.
10. Krishnamurthy A, Khardekar R, McMains S, Haller K, Elber G. Performing efficient NURBS modeling operations on the GPU. IEEE Trans Visual Comput Graphics. 2009;15(4):530–43. doi:10.1109/TVCG.2009.29.

11. Kong LB, Cheong CF. Design, fabrication and measurement of ultra-precision micro-structured freeform surfaces. Comput Ind Eng. 2011;61(1):216–25.
12. Neo DWK, Kumar AS, Rahman M. ‘CAx technologies for hybrid fast tool/slow slide servo diamond turning of freeform surface. In: Proceedings of the institution of mechanical engineers, Part B, J. Engineering Manufacture, 2016; 230(8):1465–79.

# Chapter 4
# Development of Hybrid FTS/SSS Diamond Turning

This chapter presents a hybrid fast tool and slow slide servo (FTS/SSS) diamond turning process with layered tool trajectories to address the limited FTS travel length and low bandwidth in the existing machine tool.

Generally, a hybrid freeform surface has a large sag height and may have several freeform surfaces. Thus, it is necessary for an ultraprecision machine to have large working depth (Z-axis) and high bandwidth system to machine the hybrid freeform surface. In general, the FTS technology equipped with a high bandwidth system has shorter working depths. On the other hand, SSS technology has larger travel length but it is a low bandwidth system, resulting in slower spindle speeds which may degrade the machined surface quality. The working travel length and the bandwidth are two separate performance parameters which cannot be optimized simultaneously [1]. Hence, one way to overcome this problem is to modify or upgrade the FTS system with larger travel length. Achieving larger travel length may take a lot of efforts analyzing the existing FTS system and modifying it to fulfill the surface height requirement.

## 4.1 Principle of Layered Tool Trajectory

In contrast to standard FTS diamond turning which employs three axes as explained in previous Chap. 3, the proposed layered tool trajectory method is incorporating an existing Z-axis, as shown in Fig. 4.1, in the SSS system to perform a hybrid FTS/SSS diamond turning process.

Figure 4.2a illustrates that C-axis rotational movement (spindle) stops at a trajectory point where the FTS tool reaches its maximum travel stroke, $W_{max}$ (Position (ii))

Some content of this chapter has been reproduced with permission from [1, 3].



D.W.K. Neo, *Ultraprecision Machining of Hybrid Freeform Surfaces Using Multiple-Axis Diamond Turning*, Springer Theses,
DOI 10.1007/978-981-10-4083-2_4

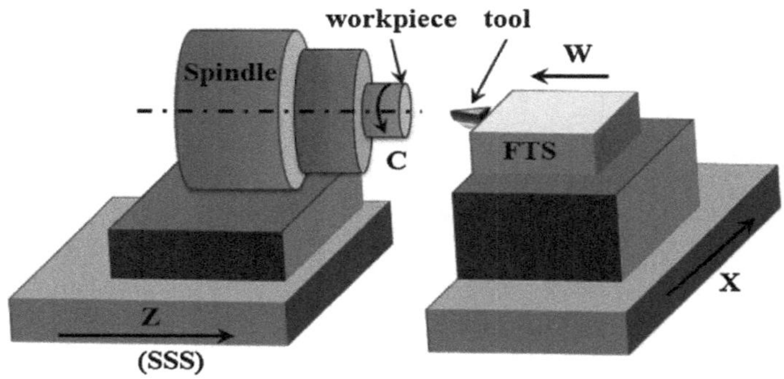


**Fig. 4.1** Schematic diagram of a hybrid FTS/SSS turning machine

during the cutting process. At the same time, the workpiece would retract downwards by the Z-axis motor controller at the same distance of $W_{max}$. Then, the FTS also retracts its tool by $W_{max}$ (Position (iii)) and resumes its cutting process (Position (iv)). The repetition of above cycle shall be continued until the tool reaches the end of spiral tool trajectory. Hence, the travel of FTS from Eq. (3.6) shall be re-arranged as:

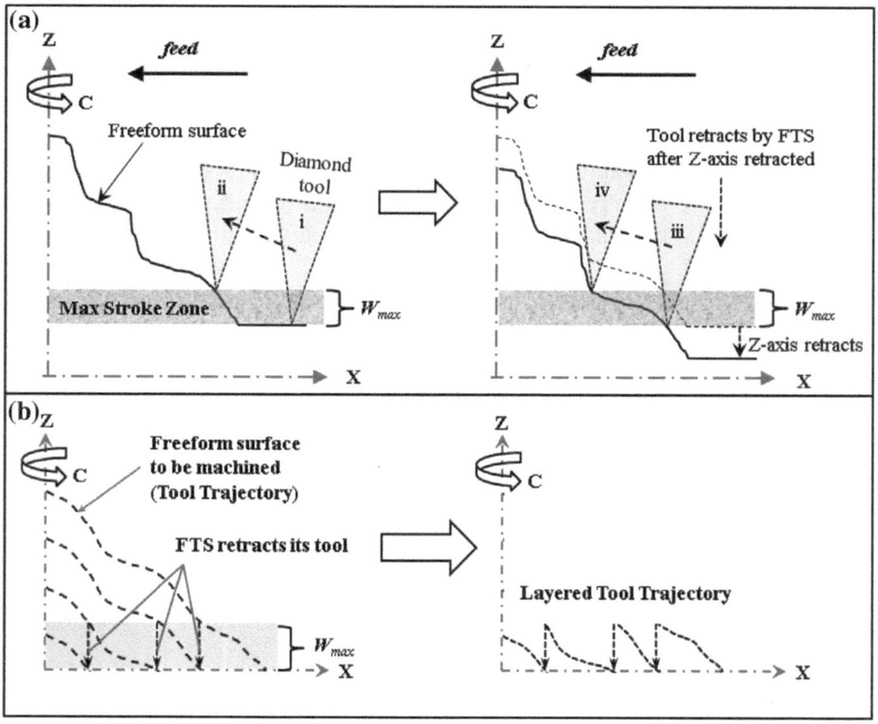


**Fig. 4.2** **a** Principle mechanisms of layered tool trajectory, and **b** the formation of layered tool trajectory. Figure reproduced from [1]

$$W_i = f(\rho, \theta) - W_{\max} \times i, \quad i = 0, 1, 2\ldots \tag{4.1}$$

where $W_{max}$ and $i$ are the maximum travel length of FTS and $i$th Z-axis retraction, respectively. Figure 4.2b demonstrates that the developed layered tool trajectory is a modification of an original tool trajectory with several number of Z-axis retractions.

## 4.2 Layered Tool Trajectory Control

Figure 4.3 describes the process flowchart for generation tool trajectory points of the hybrid FTS/SSS process. Firstly, the control points are generated based on constant-angle method, where the constant angular interval, $\Delta\theta$ is explained in previous Eq. (3.7).

Secondly, a tool nose radius compensation has to be implemented to overcome any overcutting into the surface, as shown in Fig. 4.4 [2]. The compensated depth $E_z$ can be defined as:

$$E_z = \left(\frac{1}{\cos\beta} - 1\right) \times r_t \tag{4.2}$$

where $\beta$ and $r_t$ are an inclined slope at the tool tip (cutting point) and a tip radius of the cutting tool, respectively. Thus, the compensated FTS travel length $W_c$ shall be:

$$W_c = W^* + E_z \tag{4.3}$$

where $W^*$ is the FTS travel length.

Next, the entry/exit point $P_i$ has to be properly defined for an accurate layered tool trajectory. Figure 4.5 illustrates that there would be an overcutting problem when the Point $P_i$ is not well assigned. This is due to dynamic response of C-axis motion. Thus, it is desired to have a sufficient dynamic response recovery in the C-axis motion preventing unnecessary overcuts [1]. This can be done using a forward-tracking approach, as described in Fig. 4.6, and to detect any overcut which may occur during the dynamic response recovery of C-axis motion at the point $P_i$. There will be no detection of overcutting, if:

$$z_{i+j} \le z_i, \quad j \in \left\{1, \frac{\theta_f}{\Delta\theta}\right\} \tag{4.4}$$

where $j$ is the forward-tracking position to detect for any overcut and $\theta_f$ is the clearance angle for sufficient dynamic response recovery of C-axis motion. When there is no detection of overcut, the Point $P_i^*$ with free of overcuts is the Point $P_i$.

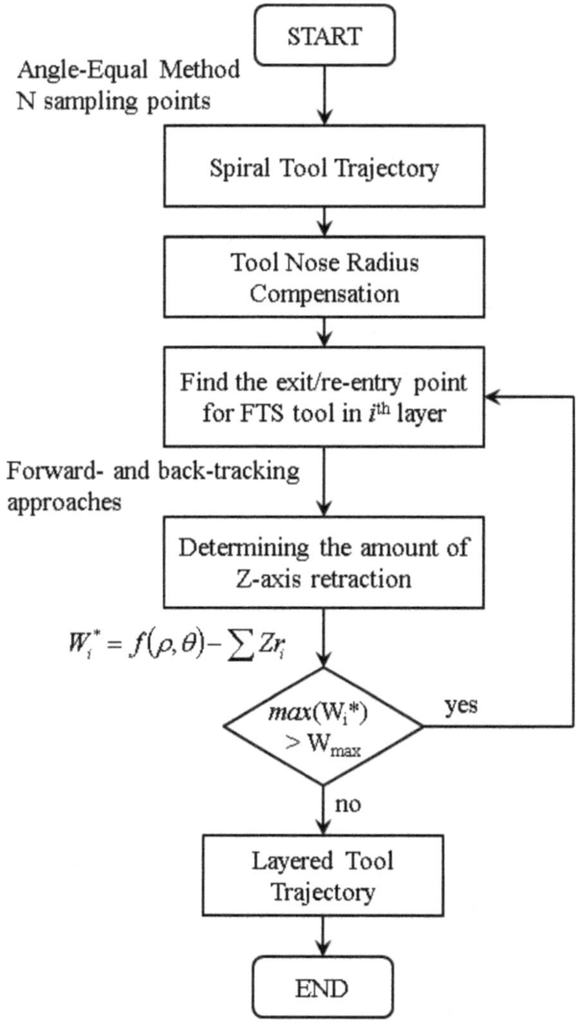


**Fig. 4.3** Process flowchart for generating layered tool trajectory

Otherwise, the backtracking approach has to be implemented to move to the previous spiral point $P_{i-k}$ (where $k$ = 1, 2 …) and repeat the forward-tracking process, as in Eq. (4.4). This back-tracking process shall be repeated until $P_i^*$ has been defined.

After the Point $P_i$ has been well defined, Z-axis retraction $Zr_i$ would be the next. Figure 4.7a demonstrates that there would be an undercutting issue in the negative region in Z-axis due to large $Zr_i$. When $Zr_i$ is well defined, these undercuts can be eliminated, as illustrated in Fig. 4.7b. Hence, $Zr_i$, as shown in Fig. 4.7, can be pre-defined with the maximum Z-axis boundary $Zb_{max}$ which lies in a surrounding radius $\rho(P_i^*)$, as follows:

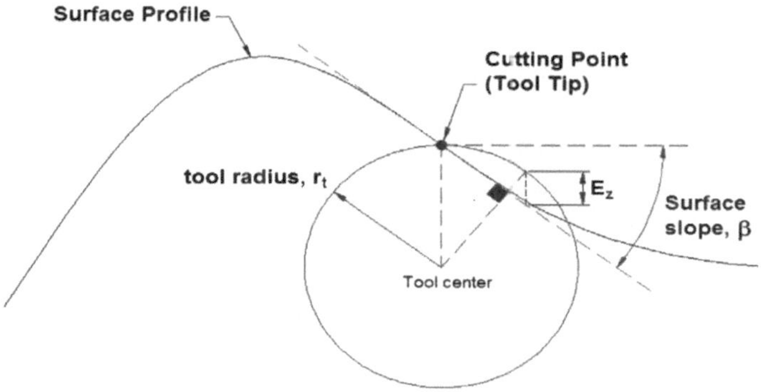


**Fig. 4.4** Overcutting of surface without tip radius compensation. Figure reproduced from [1]

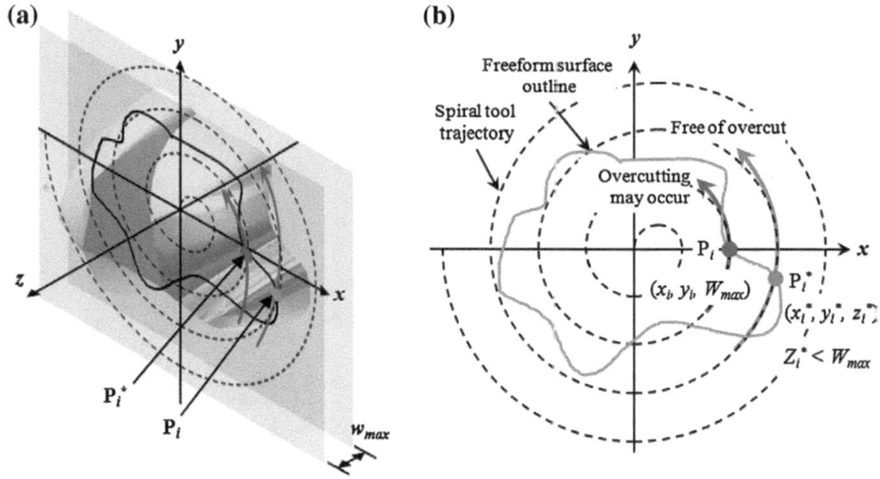


**Fig. 4.5** Exit and re-entry points on upper limit of FTS stroke zone, **a** 3D projection and **b** cross-sectional projection at $W_{max}$

$$
\begin{aligned}
Z_{\rho(Pi^*)} &= \arg\min\left[f\left(\rho\left(P_i^*\right), \theta\right)\right], \text{for } \theta \in [0^\circ, 360^\circ] \text{ in uniform step of } \Delta\theta \\
Zb_{\max} &= Z_i^* - Z_{\rho(Pi^*)} \\
Zr_i &= Z_i^* - Zb_{\max}
\end{aligned}
\tag{4.5}
$$

where $Z_i^*$ and $Z_{\rho(P_i^*)}$ are the z-axis of point $P_i^*$ and the minimum intersected height of surface and cylindrical region around $\rho(P_i^*)$. Thus, the final FTS travel length $W_i^*$ is re-written as (Fig. 4.8):

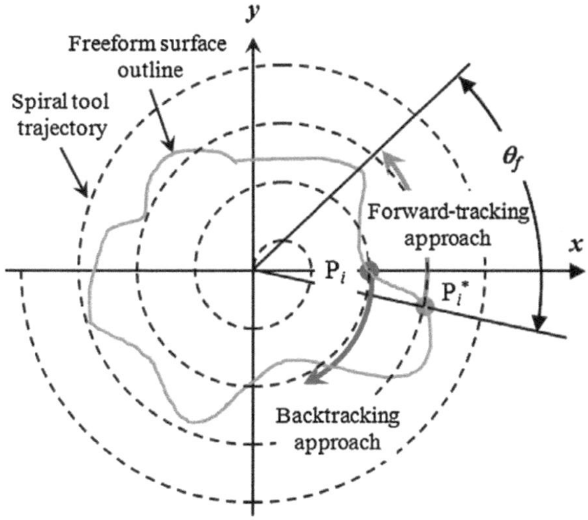


**Fig. 4.6** Forward- and back-tracking approaches for the detection of overcuts

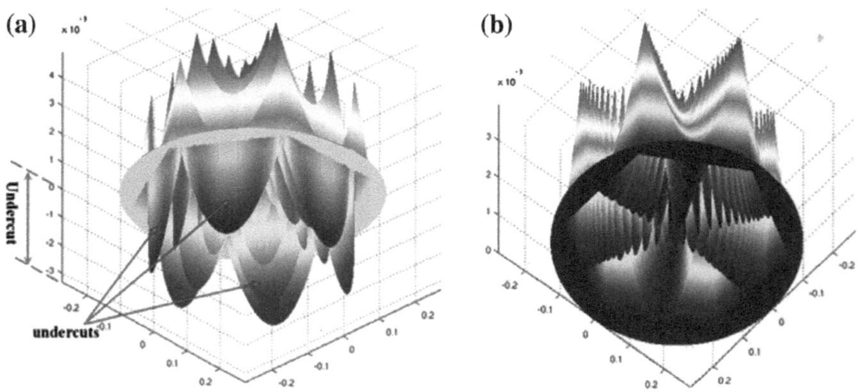


**Fig. 4.7** Influence of Z-axis retraction, **a** over-retraction, and **b** correct retraction. Figure reproduced from [1]

$$W_i^* = f(\rho, \theta) - \sum Zr_i, \quad \text{where} \quad 0 < Zr_i < Z_i^*. \tag{4.6}$$

Lastly, the layered tool trajectory is generated and is post-processed into a NC file for the machining of hybrid freeform surface using hybrid FTS/SSS diamond turning process.

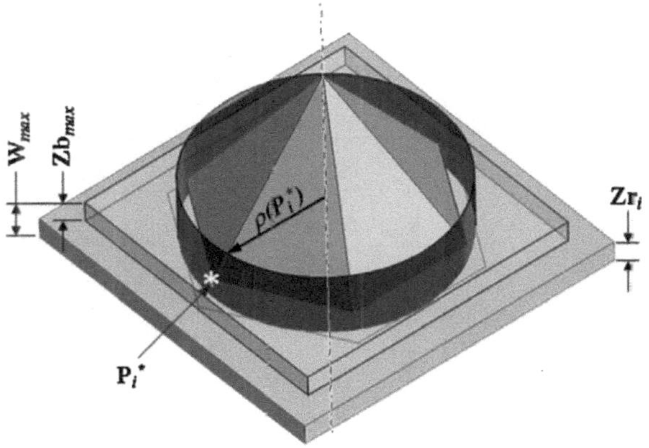


**Fig. 4.8** Schematic diagram for determining Z-retraction. Figure reproduced from [3]

## 4.3 Experimental Validations

Several experiments were conducted to validate with the fabrication of a single hexagonal micro prism, as illustrated in Fig. 4.9. The $w$-value for the tool trajectory with respect to cutting point ($x_i$, $y_i$) can be described as:

$$w_i = \begin{cases} ht - \rho_i \cos\left|\gamma_i - \frac{\theta_c}{2}\right| \tan\alpha & , if\ \rho_i \cos\left|\gamma_i - \frac{\theta_c}{2}\right| \le r_{hex}\cos\left(\frac{\pi-\theta_c}{2}\right) \\ 0 & , \text{otherwise} \end{cases}$$
$$\gamma_i = MOD\left[\left|\text{atan2}\left(\frac{y_i}{x_i}\right)\right|, \theta_c\right] \qquad (4.7)$$
$$\theta_c = 2\pi/N_s$$

where $ht$, $r_{hex}$ and $\theta_c$ are the vortex height, circumscribed radius and central angle of hexagonal micro-prism, respectively, $N_s$ is the number of sides in a regular hexagonal, $\gamma_i$ is the angle between $r_{hex}$ and $\rho_i$, and atan2 is the four-quadrant inverse-tangent function which returns the results in a range (−π, π). Table 4.1 shows the cutting parameters which are selected for fabricating a single hexagonal micro-prism. Figure 4.10 shows the simulation of layered tool trajectory for machining micro prism and a photographic view of machined micro prism. The machined hexagonal micro-prism is found to have a slight profile distortion. This was due to an error of tool de-centering which was about Ø5 μm.

A white light interferometer has been employed for measuring the height and surface roughness of fabricated micro prism. Figure 4.11 shows that the measured result for the vortex height was 25.09 μm and an error of 0.66 μm (4.56%) was calculated as compared to the designed height of 25.75 μm. This result is lesser than 1.0 μm which further validates the credibility the developed layer tool trajectory. Figure 4.12 illustrates that the measured average surface roughness

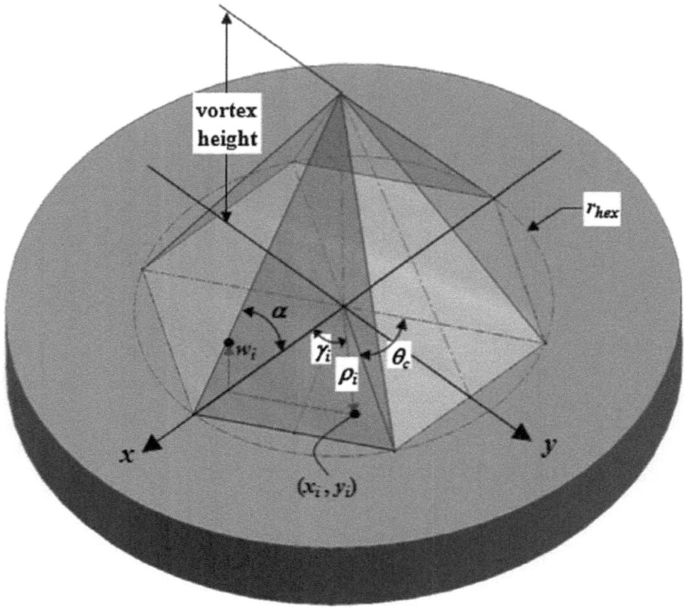


**Fig. 4.9** Schematic diagram for calculation of w-values in the micro prism

**Table 4.1** Fabrication parameters for hexagonal micro-prism

| | |
|---|---|
| Machining conditions | Feedrate: 1 μm/rev<br>Spindle speed: 50 rpm<br>Sampling number, $N_p$: 360<br>$W_{max}$: 5 μm |
| Cutting tool | SCD with 0.2 mm tool nose radius |
| Workpiece | Brass, Ø6 mm |
| Hexagonal micro-prism | Cyclic hexagonal base with circumscribed radius $r_{hex}$ of 0.25 mm |
| Vortex height, *ht* | 25.75 μm |

Table reproduced from [1]

(Ra) of a facet was about 195 nm. This surface roughness could be further improve by increasing the sampling points $N_p$, as explained in the previous Sect. 3.2. On top of that, the error in the vortex height and relatively high surface roughness were also most likely caused by the dynamic response of Z-axis motion. This dynamic response would lead to positioning error for each Z-axis retraction and cause $P_i^*$ points to overlap one another [1].

Lastly, the developed methodology of layered tool trajectory for the hybrid FTS/SSS diamond turning has been successfully exhibited the extension of effective FTS travel length to 25.75 μm which is about five folds of its maximum travel length (5 μm), without replacing/upgrading the existing FTS system.

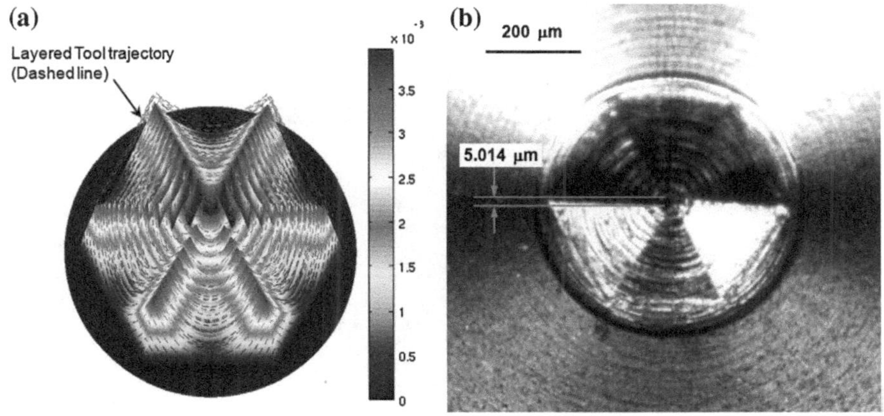


**Fig. 4.10** **a** Simulated layered tool trajectory for hexagonal micro prism and **b** fabricated hexagonal micro prism. Figure reproduced from [1]

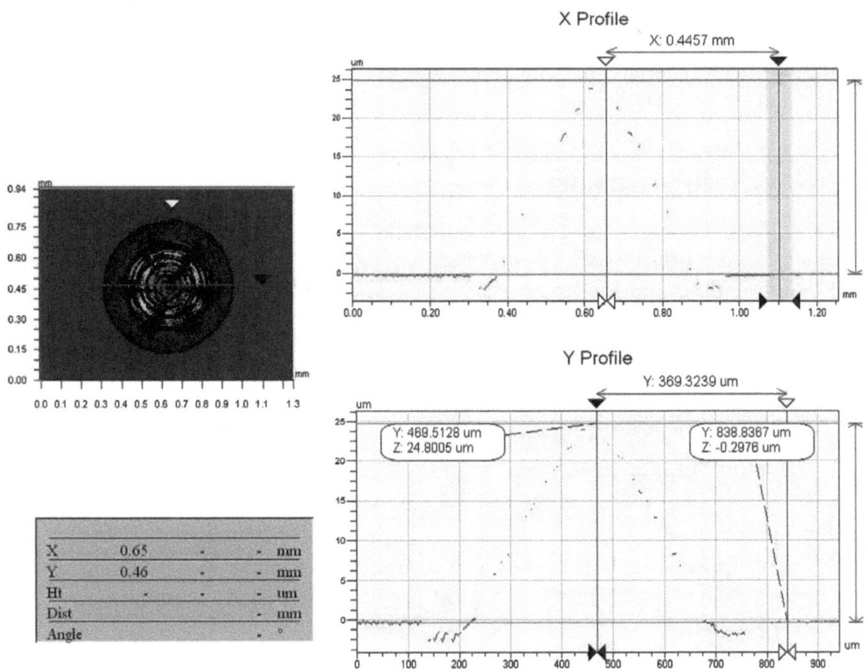


**Fig. 4.11** Overall height measurement of fabricated micro prism. Figure reproduced from [1]

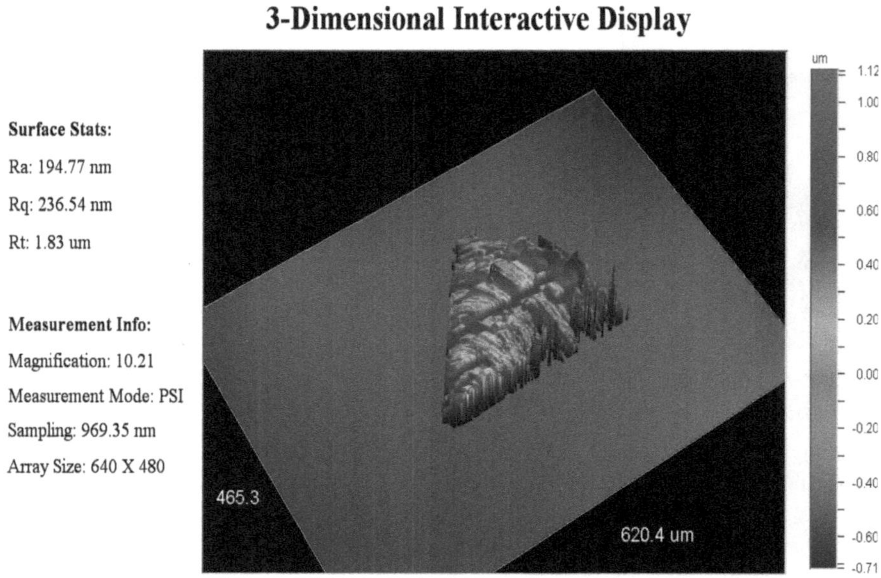


**Fig. 4.12** Surface roughness measurement on a single face of fabricated micro prism. Figure reproduced from [1]

## 4.4 Concluding Remarks

In this study, a hybrid (FTS/SSS) process has been developed to increase the FTS travel length without modifying an existing FTS system. The following conclusions can be drawn:

(i) The implementation of the layered tool trajectory methodology has demonstrated the feasibility study of fabricating a hexagonal micro prism with hybrid FTS/SSS diamond turning.
(ii) The height of fabricated micro prism has an error of 0.65 μm as compared to designed height requirement, which also fulfils the profile accuracy requirement of 1.0 μm.
(iii) The proposed methodology has successfully extended an effective stroke length of existing FTS system to about five folds of its maximum possible stroke length.

However, there are few areas which are required to be address for improvements. They are as follows:

(i) It is difficult to fabricate an offset micro prism (more complex hybrid freeform surface). This is due to the loss of symmetry in the complex hybrid freeform surface. Hence a novel surface generation study has to be done by incorporating more machining axes to overcome the loss of symmetry. The details of the developed methodology are discussed in the next chapter of this dissertation.

(ii) Poor surface roughness is one of the problems faced in the diamond turning application. This is most likely due to the large cutting linearization errors resulted from the poor selection of sampling points per revolution, $N_p$. Thus, a surface analytical model is necessary to be derived for optimizing this cutting linearization error. The developed surface analytical model is explained in details in Chap. 6 of this dissertation.

(iii) Assignments for W- and Z-axes in the layered tool trajectory methodology are only designed for micro prism but not for any freeform feature. Hence, a proper segregation technique for any freeform feature has to be established so that the fabrication of such freeform surfaces can be executed by using hybrid FTS/SSS process. This is explained in detail in Chap. 7.

## References


1. Neo WK, Kumar AS, Rahman M. A novel method for layered tool path generation in the fast tool servo diamond turning of non-circular microstructural surfaces. Proc Inst Mech Eng Part B, J Eng Manuf. 2013; 227(2):210–9.
2. Hsu WY, Chen FZ, Kuo CH, Cheng YC, Lin WL. Fabrication of micro aspherical lens array using fast tool servo with tool path compensation. In: ASPE proceedings, annual meetings.
3. Senthil Kumar A, Aravind Raghavendra MR, Neo WK, Rahman M. Fast and fine tool servo for ultraprecision machining in comprehensive materials processing. In: Rahman M, editor, vol. 11: Advanced Machining Technologies, Elsevier Ltd. 2014; p. 61–88. doi:10.1016/B978-0-08-096532-1.01104-3.

# Chapter 5
# Novel Surface Generation of Complex Hybrid Freeform Surfaces

This chapter presents a novel surface generation technique with multiple machining axes to address the loss of symmetrical axes in the machining of complex hybrid freeform surfaces. An automated Guilloche machining technique (AGMT) has been developed in the multiple-axis diamond turning machine to address the loss of symmetry. Fresnel lens array is one of these complex hybrid freeform surfaces, which consists of several elements of Fresnel lenses arranged in a rectangular or hexagonal layout. In general such array of Fresnel lenses is manufactured individually and assembled, which cannot be achieved in a single process. This is due to the increase of complexity associated with a loss of symmetry of the surface [1]. This also leads to a loss of accuracy in assembly process and unnecessary parting lines between the Fresnel elements and deteriorates the optical outputs of Fresnel lens arrays.

The proposed novel surface generation employs four machining axes and moves a diamond tool similar to a hand-operated Guilloche machine. Section 5.1 presents the novel surface generation for AGMT in machining of Fresnel lens array. Following by, Sect. 5.2 discusses two analytical approaches to evaluate the critical machining parameters for fabricating accurately machined Fresnel lens arrays. In additions, these two approaches were further validated experimentally for their creditability. Section 5.3 presents the concluding remarks which summarize the results from this study.

## 5.1 Novel Surface Generation for Automated Guilloche Machining Technique

The proposed automated Guilloche machining technique (AGMT) employs an ultraprecision machine tool with three translation motions ($x$, $y$ and $z$-axes) and one rotational motion ($c$-axis). A tip of the cutting tool, as illustrated in Fig. 5.1a, moves

---

Some content of this chapter has been reproduced with permission from [2].



D.W.K. Neo, *Ultraprecision Machining of Hybrid Freeform Surfaces Using Multiple-Axis Diamond Turning*, Springer Theses,
DOI 10.1007/978-981-10-4083-2_5

along a circular arc of radius $r_c$ using x- and y-machine axes, but at an offset distance $\rho$ from spindle axis. This resulted in a circular cutting action which circumscribes around a small circle of radius $\rho$ and is also analogous to a Guilloche trajectory [2]. Simultaneously, the workpiece or machine spindle rotates in the same rotational direction (clockwise) of Guilloche tool trajectory. In this manner, the tool only cuts the workpiece in a direction tangentially to that smaller circle, as explained in Fig. 5.1b, behaving as a conventional facing operation in the offset position.

A complete Guilloche cutting trajectory for a circular Fresnel lens is in the form of an Archimedes spiral, which can be also considered as the same as those trajectories for FTS/SSS diamond turning processes, as explained in previous Chap. 3. Hence, Eqs. (3.6–3.7) can be re-written as:

$$Cx_i = \rho_i \sin t_i \tag{5.1}$$

$$Cy_i = \rho_i \cos t_i \tag{5.2}$$

$$t_i = i \times \left(2\pi/N_p\right), \quad \text{for } i \in (0, N_g) \tag{5.3}$$

where $Cx_i$ and $Cy_i$ are the coordinates of the cutting point $P_i$, $\rho$ and $t$ are the radial and rotational positions for the workpiece or spindle, respectively, $i$ is the $i$th angular position, $N_p$ and $N_g$ are the number of angular divisions per revolution and total number of Guilloche trajectory points to complete a cutting trajectory in each Fresnel lens. However, the tool trajectory for any offset Fresnel lens would behave differently as compared to that of centre one. Hence, the Guilloche cutting trajectory of offset Fresnel lenses have to be translated in a manner such that the cutting tool moves along its translated tool trajectory. Firstly, the tool moves from a cutting point $P_i$ to next $P_{i+1}$ and is represented by:

$$P_{i+1} = (x_{i+1}, y_{i+1}, z_{i+1}, \theta_{i+1}) \tag{5.4}$$

where $\theta_{i+1}$ is a rotational coordinate of workpiece/spindle, $x_{i+1}$ and $y_{i+1}$ are the translations of Eqs. (5.1) and (5.2), respectively, with respect to $\rho_i$ and $r_c$:

$$\theta_{i+1} = \theta_0 - t_i \tag{5.5}$$

$$Xc_{i+1} = r_c \sin \theta_{i+1} \tag{5.6}$$

$$Yc_{i+1} = r_c \cos \theta_{i+1} \tag{5.7}$$

$$x_{i+1} = Xc_{i+1} + \rho_{i+1} = r_c \sin \theta_{i+1} + \rho_{i+1} \tag{5.8}$$

$$y_{i+1} = Yc_{i+1} = r_c \cos \theta_{i+1} \tag{5.9}$$

where $Xc$ and $Yc$ are the centre points of a Fresnel lens.

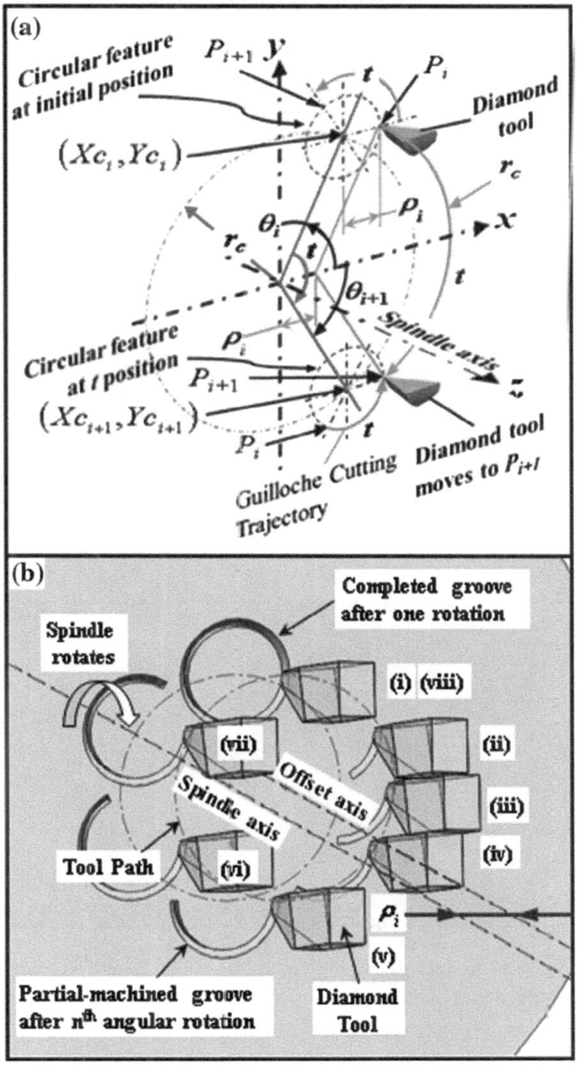


**Fig. 5.1** Principle of AGMT; **a** schematic diagram and **b** simulation of cutting kinematics. Figure reproduced from [2]

## 5.2 Experimental Validations

In this study, several experiments have been conducted to machine 3 × 3 circular Fresnel lenses array, as illustrated in Fig. 5.2, to validate the feasibility of the developed automated Guilloche machining technique. The surface generation for the Guilloche cutting trajectory was processed by implementing the above methodologies into the developed SolidWorks-API, as discussed previously in Chap. 3. Figure 5.3 shows that the cutting points were successfully mapped onto the surface of Fresnel lenses.

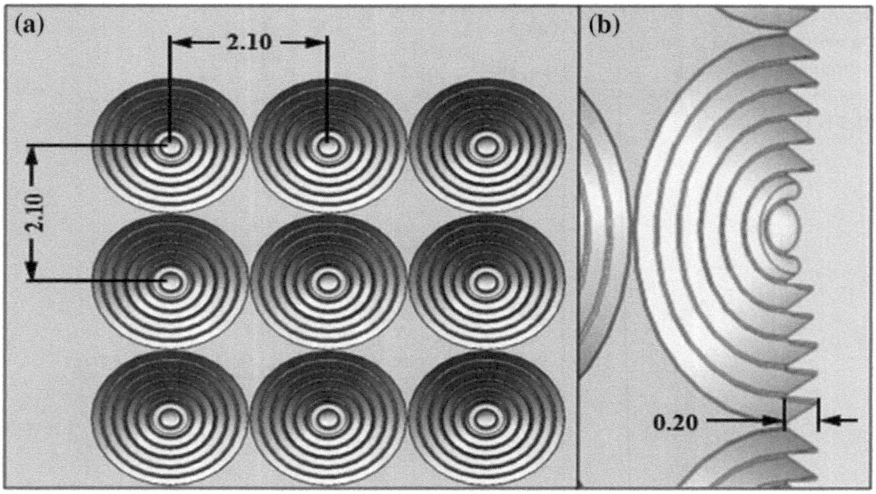


**Fig. 5.2** A CAD model for Fresnel lens array; **a** in rectangular layout **b** half-sectional view

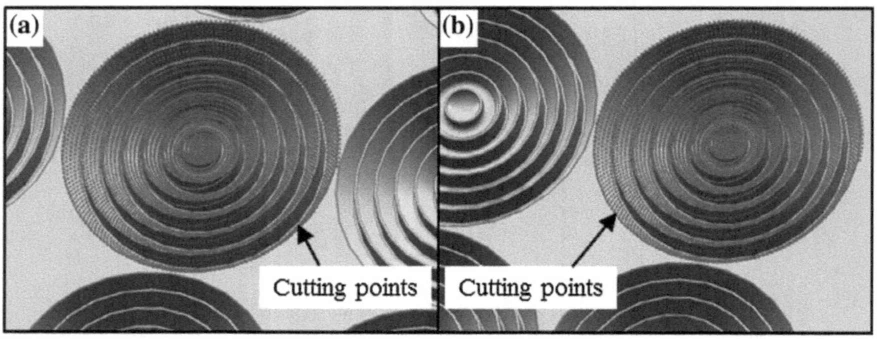


**Fig. 5.3** Successful mapping of spiral points using the developed SolidWorks-API system; **a** on central Fresnel lens, and **b** on an offset Fresnel lens

## *5.2.1 Critical Machining Parameters*

In order to generate accurate cutting points, the critical machining parameters have to be defined before they could be further translated into NC-codes. In this study, there are two critical parameters, namely feedrate $\Delta\rho_{cr}$ and angular pitch $\Delta t_{cr}$.

### 5.2.1.1 Critical Feedrate $\Delta\rho_{cr}$

A cutting residual error $E_\rho$, as illustrated in Fig. 5.4, is cusp height of two corresponding tool marks/tool nose profiles along the feed direction on a curved surface.

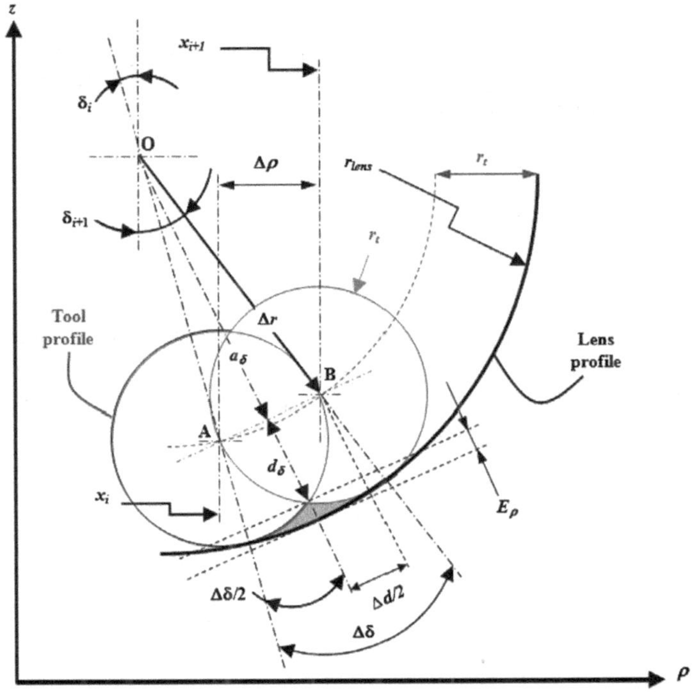


**Fig. 5.4** Schematic diagram for evaluating $E_p$ along the feed direction. Figure reproduced from [2]

This $E_\rho$ is strongly depending on the proper selection of feedrate $\Delta\rho$ and can be defined as [2]:

$$
\begin{aligned}
x_{i+1} &= x_i + \Delta\rho \\
\delta_{i+1} &= \sin^{-1}\left(\frac{x_{i+1}}{\Delta r}\right), \quad \text{where } \Delta r = r_{lens} - r_t \\
a_\delta &= \Delta r \cos(\Delta\delta/2) \quad \text{where } \Delta\delta/2 = (\delta_{i+1} - \delta_i)/2 \\
d_\delta &= \sqrt{r_t^2 - \left(\frac{\Delta d}{2}\right)^2}, \quad \text{where } \frac{\Delta d}{2} = \frac{1}{2}\Delta r \sin\left(\frac{\Delta\delta}{2}\right) \\
E_\rho &= r_{lens} - a_\delta - d_\delta
\end{aligned}
\tag{5.10}
$$

where $\delta$ is the angular position of tool profiles with respect to the centre of lens curvature at point $O$, $\Delta\delta$ is the angle between two tool profiles, $\Delta r$ is the radii difference between the lens curvature $r_{lens}$ and tool nose radius $r_t$, $a_\delta$ is the apothem of the triangle $AOB$, $\Delta d$ is the distance $AB$, and $d_\delta$ is the Euclidean distance from the mid-point of $AB$ to the tip of cusp.

Table 5.1 describes a list of lens curvatures for each Fresnel zone. From the simulated results of several feedrates for a tool nose radius of 10 μm as shown in Fig. 5.5, it can be seen that the selection of feedrates should be lesser than the critical feedrate $\Delta\rho_{cr}$ of 0.001 mm/rev. Otherwise, $E_p$ would be failed to meet the targeted error of 0.1 μm if the feedrates are larger than $\Delta\rho_{cr}$.

### 5.2.1.2 Critical Angular Pitch $\Delta t_{cr}$

In contrast to a circular interpolation from a CNC system, the tool trajectory of AGMT generates based on several linear segments to form a polygonal arc, instead of circular arc. Thus, a resulted Sagitta error $h_{err}$, as illustrated in Fig. 5.6, is the sag height difference between polygonal and circular arcs. Hence, it necessary to minimize this $h_{err}$ by determining the critical angular pitch $\Delta t_{cr}$:

**Table 5.1** List of lens curvatures in each Fresnel zone

| Parameters | Zones 1–6 | Zone 7 | Zone 8 |
|---|---|---|---|
| $r_{lens}$ (mm) | 50.0 | 0.05 | 0.15 |

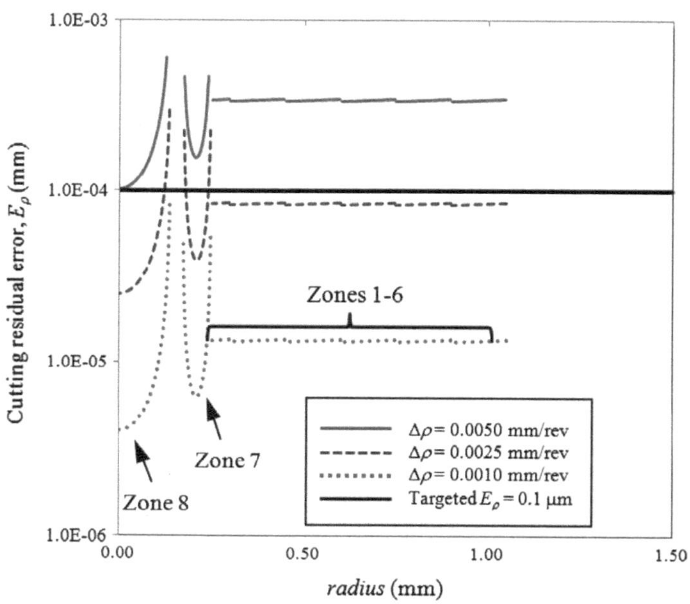


**Fig. 5.5** Simulated results of $E_p$ with different feedrates on individual Fresnel zones

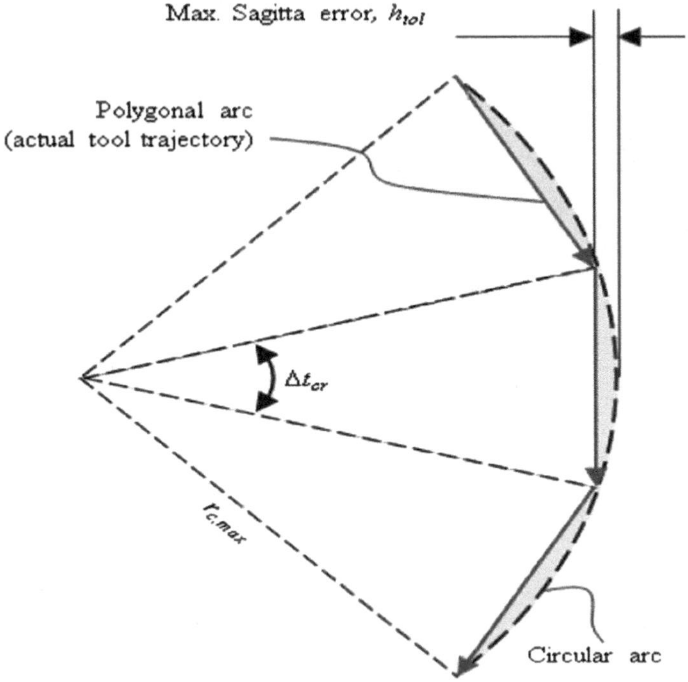


**Fig. 5.6** Schematic diagram for calculating $\Delta t_{cr}$

$$
\begin{aligned}
h_{tol} &= r_{c,\max}\left(1 - \cos\frac{\Delta t_{cr}}{2}\right) \\
\Delta t_{cr} &= 2\cos^{-1}\left(1 - \frac{h_{tol}}{r_{c,\max}}\right)
\end{aligned}
\tag{5.11}
$$

where $h_{tol}$ is the targeted $h_{err}$ and $r_{c,max}$ is maximum arc radius of the tool trajectory. From Fig. 5.7, it can be seen that $\Delta t$ decreases gradually in order to meet the targeted $h_{tol}$ of 0.1 μm when $r_c$ becomes further away from the centre of workpiece/spindle. Hence, the critical value $\Delta t_{cr}$ for $r_{c,\max}$ of 4.455 mm should be lower than 0.768°, or else it would be failed to generate an accurate tool trajectory.

#### 5.2.1.3 Cutting Experiments and Results

All the experiments were conducted by AGMT process using a 4-axis ultraprecision machine. Based on the simulated results from the above approaches, the machining parameters (Table 5.2) were selected in machining 3 × 3 Fresnel lens array to meet the targeted $h_{tol}$ of 0.1 μm. These machined Fresnel lenses (as shown in Fig. 5.8) were measured using an Olympus LEXT OLS4000 3D Measuring Laser

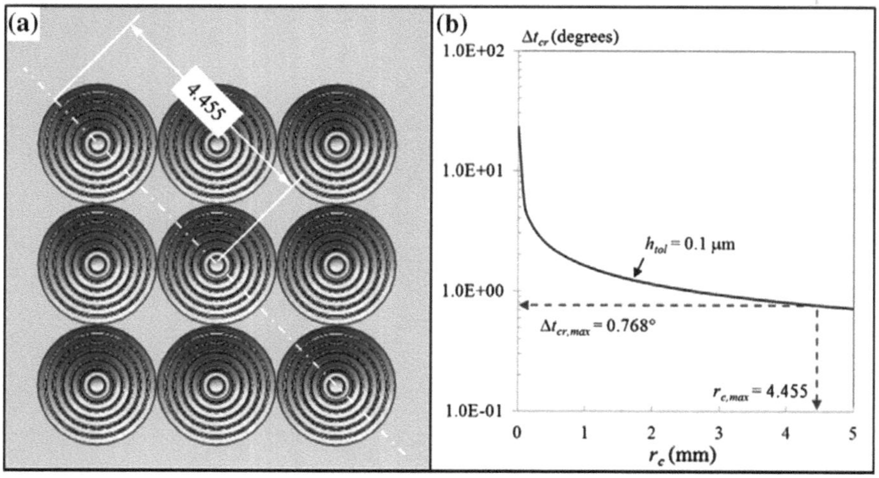


**Fig. 5.7** Simulated results of $\Delta t_{cr}$ for targeted $h_{tol}$ of 0.1 μm

**Table 5.2** Machining parameters for machining 3 × 3 Fresnel lens array

| Machining parameters | Value | Units |
|---|---|---|
| Spindle speed | 50.0 | rpm |
| Feed, $\Delta\rho$ | $0.001 = \Delta\rho_{cr}$ | mm/rev |
| Angular pitching, $\Delta t$ | $0.75 < \Delta t_{cr}$ | ° |
| Depth of cut | 0.005–0.001 | mm |
| Workpiece | AA6061-T6 | – |
| Tool | Mono-crystalline diamond, 10 μm nose radius, 40° front clearance | – |

Microscope with a confocal optical system. 3D measured profile data were further post-processed using MATLAB for surface characterization.

Figure 5.9a, b show the measured 3D profile of the central circular Fresnel lens and its comparison results between the ideal and the measured profiles, respectively. It can be observed that the results of cutting residual error $E_\rho$ were found to be about 0.085 μm which were lesser than the required accuracy of 0.1 μm, as shown in Fig. 5.9c.

Figure 5.10a, b show the measured 3D profile and the comparison results between the ideal and measured profiles for an offset circular Fresnel lens, respectively. The results of $E_\rho$ were found to be about 0.080 μm which were also lesser than the required accuracy of 0.1 μm, as shown in Fig. 5.10c.

The Sagitta error $h_{err}$ for the central Fresnel lens is not evaluated as the arcs of cutting trajectory are revolved about the spindle axis and the cutting process

**Fig. 5.8** Photographic view of a machined 3 × 3 Fresnel lens array

behaves as a conventional facing operation in the turning process. Hence, there is no formation of linear segments along the arcs and $h_{err}$ would not occur in the central Fresnel lens.

Thus, these results further validate that the selected machining parameter, the feedrate $\Delta\rho$, for cutting the circular Fresnel lens array is achieving the cutting residual error $E_\rho$ lesser than 0.1 μm.

Figure 5.11a illustrated a selected region of an offset Fresnel lens was been replicated using MATLAB for evaluating the Sagitta errors $h_{err}$. Its replicated results for the selected region were demonstrated in Fig. 5.11b. Figure 5.11c, d show the evaluated results of $h_{err}$ were plotted in the 3-dimensional and 2-dimensional, respectively, and were found to be lesser than the required accuracy of 0.1 μm. These results further validate that the selected second parameter, the angular pitching $\Delta t$ for machining circular Fresnel lens array is also achieving the accuracies lesser than the $h_{tol}$ of 0.1 μm.

Lastly, these experimental results have validated the credibility of the developed novel surface generation for automated Guilloche machining technique to fabricate a circular Fresnel lens array in a single process, accurately.

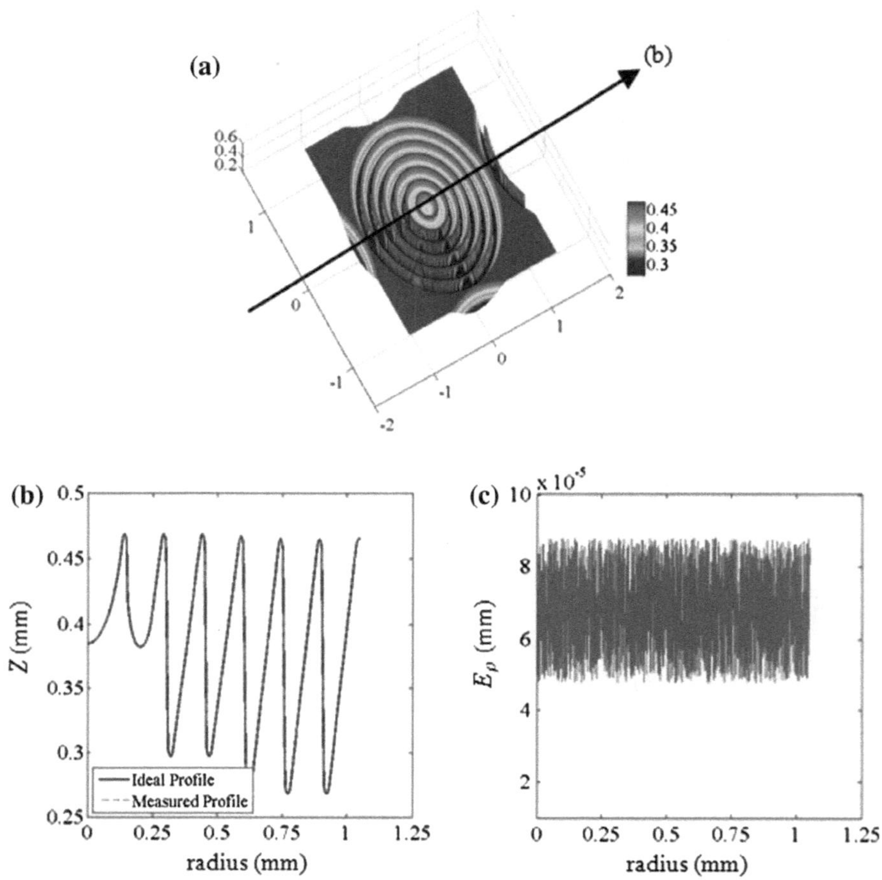


**Fig. 5.9** Measured $E_\rho$ of a machined circular Fresnel lens (central lens)

## 5.3 Concluding Remarks

In the present study, the automated Guilloche machining technique (AGMT) has been successfully developed to fabricate a 3 × 3 circular Fresnel lens array without the needs of having multiple setups. The following conclusions can be drawn:

(i) Two critical machining parameters, namely critical feedrate $\Delta\rho_{cr}$ and angular pitching $\Delta t_{cr}$ have been successfully defined to achieve accurate Guilloche cutting trajectory. These were well validated with the machining of a 3 × 3 rectangular array of circular Fresnel lenses.

(ii) The surface characterization for machined Fresnel lenses has demonstrated that the cutting residual errors $E_\rho$ and Sagitta errors $h_{err}$ were within the targeted requirements.

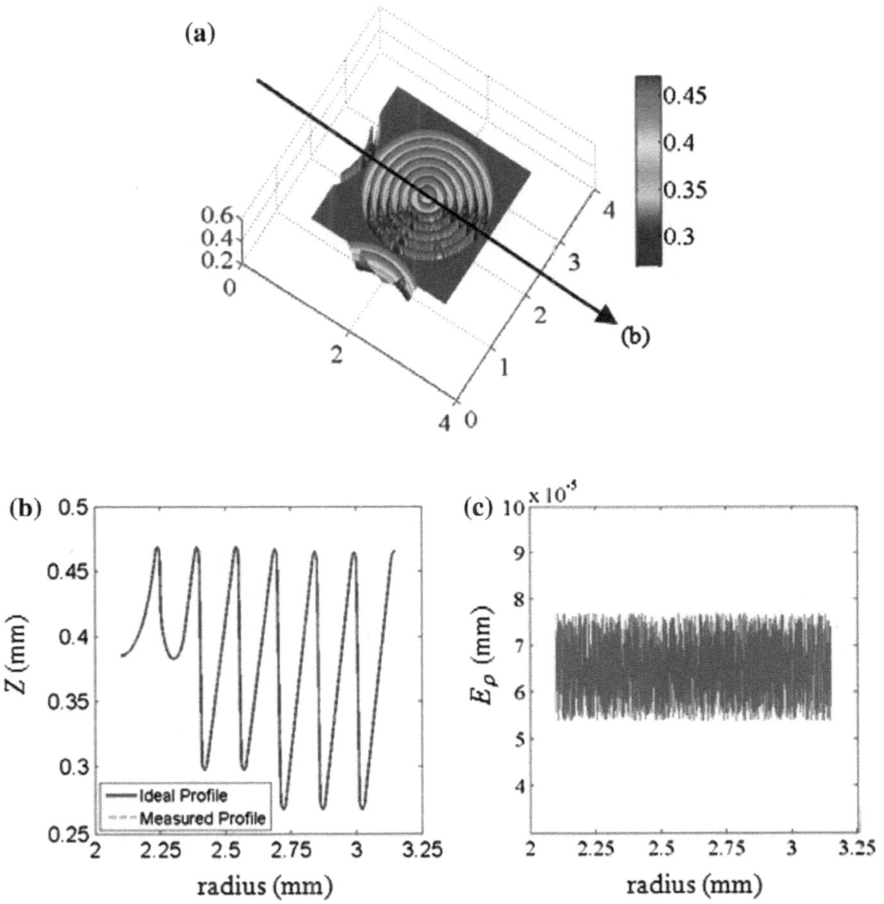


**Fig. 5.10** Measured $E_\rho$ of an offset Fresnel lens

(iii) Lastly, these results not only validate the credibility of the developed multiple-axis ultraprecision machining process for accurate complex hybrid freeform surfaces, but also offer new promising solution in machining freeform surfaces with complex curvatures.

Although the critical machining parameters have been successfully determined with the above methodologies, there is still a room for improvement for optimizing the profile accuracy. This analysis considers only the errors in the XY direction (two-dimensional), but not on the Z-direction (three-dimensional). This makes it vulnerable for unforeseen accumulation of profile errors in the Z-direction. Hence, this issue is addressed by developing a surface analytical model which is explained in details in the next chapter.

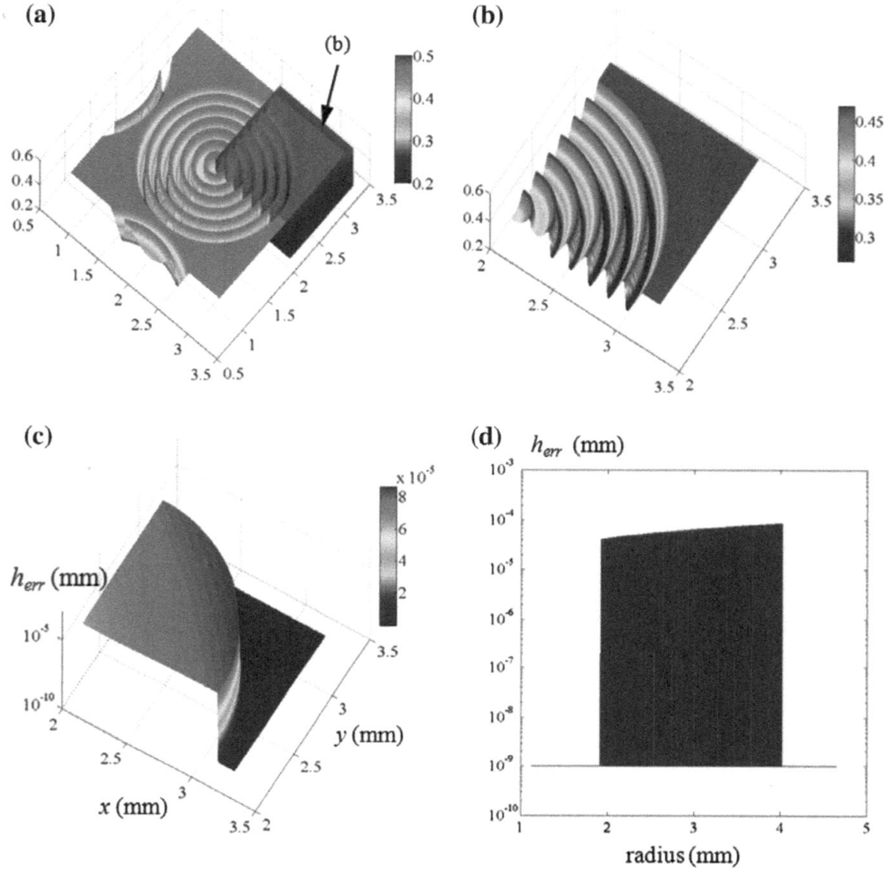


**Fig. 5.11** Measured Sagitta errors of an offset Fresnel lens

# References

1. Riemer O. Advances in ultra precision manufacturing. International symposium of the Japan society for precision engineering 2011 (ISUPEN 2011).
2. Neo DWK, Kumar AS, Rahman M. Automated Guilloche machining technique for the fabrication of polygonal Fresnel lens array. Precis Eng. 2015;41:55–62.

# Chapter 6
# Development of Surface Analytical Model for Accurate Hybrid Freeform Surfaces

This chapter presents a novel surface analytical model to determine the cutting linearization error for several cutting strategies. From the previous studies in Chap. 4, when one of the cutting strategies, constant-angle, has been employed, the surface quality of machined freeform surfaces is strongly dependent to the number of points per revolution $N_p$. This is due the induced cutting linearization errors in the spiral cutting direction, as explain earlier in Chap. 2. From literatures, it is also reported that a uniform machined surface quality could be obtained by employing another cutting strategy, constant-arc cutting strategy [1].

Notwithstanding the fact that the critical incremental arc-lengths plays a major role in achieving accurate tool trajectory [2], only few studies have been conducted for the optimization of this cutting linearization error in the FTS/SSS diamond turning. On top of that, there is no analytical study for comprehensive understanding on this cutting linearization error. Thus, lacking this important knowledge of cutting linearization error could hinder the development of FTS/STS diamond turning to fulfill the demands for ultraprecise freeform surfaces.

Two surface profiles, namely sinusoidal wave grid and microlens array are successfully machined and achieved accurate surface with proper evaluation of critical machining parameters. A hybrid constant-arc and constant-angle method for surface generation is also derived and successfully further optimized the FTS/SSS process new higher level. The results of the proposed analytical model not only provide a comprehensive understanding for the cutting linearization errors in the surface generation, but also enlighten the optimization of surface quality in the FTS/SSS machining process.

---

Some content of this chapter has been reproduced with permission from [6].



D.W.K. Neo, *Ultraprecision Machining of Hybrid Freeform Surfaces Using Multiple-Axis Diamond Turning*, Springer Theses,
DOI 10.1007/978-981-10-4083-2_6

## 6.1 Surface Generation for FTS/SSS Diamond Turning

Fast tool/slow slide servo (FTS/SSS) technology plays an important role in machining freeform surfaces for the modern optics industry. The surface accuracy is a sticking factor that demands the need for a comprehensive solution to manufacture accurate freeform surfaces efficiently. However, the analysis of cutting linearization errors in the cutting direction of surface generation has received little attention. Hence, a novel surface analytical model (SAM) has been developed to evaluate the cutting linearization errors and optimizes the number of cutting points to meet accuracy requirements. To validate the theoretical cutting linearization errors, a series of machining experiments has been conducted to machine sinusoidal wave grid and micro-lens array surfaces. The experimental results demonstrate that these surfaces have successfully achieved the targeted surface accuracy requirement of 1 μm with the implementation of the proposed model. These further credits the capability of SAM as an effective and accurate tool in improving profile accuracies and meeting accuracy requirements.

### *6.1.1 Novel Surface Analytical Model*

Surface generation is no longer just a tool path generation, but also a necessity tool for optimizing the profile accuracy with different cutting strategies. In the traditional FTS/SSS process, the machined freeform surface is often evaluated after the machining stage. This gives a high risk of having a machined surface failing to meet the profile accuracy requirements as the cutting linearization error has never been taken into account. Hence, a novel surface analytical model has been derived to address the needs for evaluating the cutting linearization error and optimizing the profile accuracy.

Two cutting strategies as described in Fig. 6.1 are constant-angle and constant-arc, which are commonly employed in the FTS/SSS process. The constant-angle method generates the control points in a manner where the number of control points for every revolution is constant. Due to the fact that the incremental arc-lengths of trajectory points at outer radii are, unfortunately, sparser than those at central radii, the surface quality of outer radii would be poorer than that at central radii. Hence, this leads to the tendency of increasing trajectory points to minimize the incremental arc-lengths for producing better surface quality at outer radii. However, the FTS/SSS systems do not permit large number of trajectory points which is larger than their bandwidths [3]. Too many control points not only reduce the spindle speed, but also leads to difficulty in cutting ductile materials which requires high cutting speed.

Thus, this longer arc-length issue could be overcome by employing the second cutting strategy known as constant-arc. This strategy not only gives a constant incremental arc-length, but also allows a uniform surface quality to be achieve.

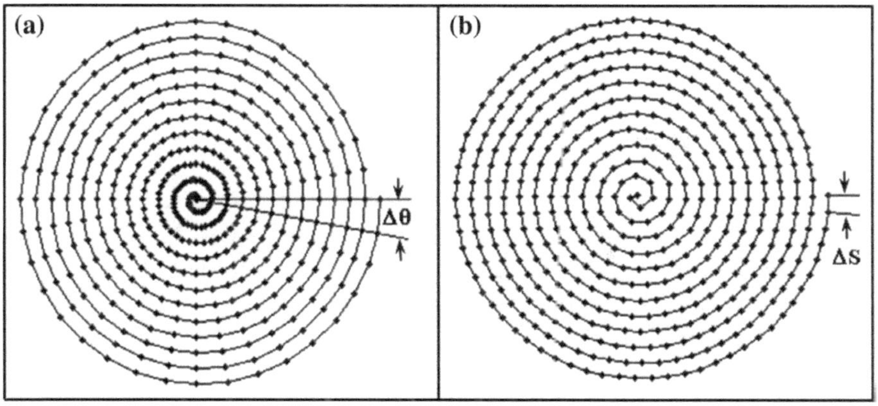


**Fig. 6.1** Cutting strategies in FTS/SSS turning, **a** constant-angle and **b** constant-arc

Nevertheless, these two cutting strategies shall be validated by the proposed surface analytical model for their machining abilities to optimize the process. The constant-angle $\Delta\theta$ and constant-arc $\Delta S$ parameters of the cutting strategies are given as:

$$\Delta\theta = 360^\circ / N_p \tag{6.1}$$

$$\Delta S = S_t / \left(N_p N_t\right) \tag{6.2}$$

where $N_p$ is the number of trajectory points per rotation and $S_t$ is the arc-length for the entire spiral tool trajectory. Both of $\Delta\theta$ and $\Delta S$ parameters play the major role in achieving surface accuracy in FTS/SSS diamond turning. Hence, it is imperative to evaluate the critical values of these parameters to meet the surface accuracy requirements.

It is known that the arc-length $S$ for Archimedes spiral usually begins from its centre and is given as [4]:

Let the radial position of diamond tool, $\rho = f_r\theta$ and

$$\begin{aligned} dS(\theta) &= \int \sqrt{\rho^2 + \left(\frac{d\rho}{d\theta}\right)^2} \cdot d\theta \\ &= \int \sqrt{(f_r\theta)^2 + f_r^2} \cdot d\theta \\ &= f_r \int \sqrt{\theta^2 + 1} \cdot d\theta \end{aligned} \tag{6.3}$$

This integral can be solved by using the integrals of irrational functions as follow [5]:

$$\int \sqrt{x^2 \pm a^2}dx = \frac{1}{2}x\sqrt{x^2 \pm a^2} \pm \frac{1}{2}a^2 \ln\left|x + \sqrt{x^2 \pm a^2}\right|$$

Hence, Eq. (6.3) can be solved as:

$$\begin{aligned} S(\theta) &= f_r\left[\frac{1}{2}\theta\sqrt{\theta^2+1}+\frac{1}{2}\ln\left|\theta+\sqrt{\theta^2+1}\right|\right] \\ &= \frac{1}{2}f_r\left[\theta\sqrt{\theta^2+1}+\ln\left|\theta+\sqrt{\theta^2+1}\right|\right] \end{aligned} \tag{6.4}$$

In the process of surface generation, the spiral tool trajectory, however, usually proceeds from the outer radius to the center of the workpiece. The details of spiral tool trajectory have been explained previously in Chap. 3 and Eq. (3.1) shall be substituted into Eq. (6.4) and can be rewritten as:

$$S(2\pi N_t-\theta)=\frac{1}{2}f_r\left[(2\pi N_t-\theta)\sqrt{(2\pi N_t-\theta)^2+1}+\ln\left|(2\pi N_t-\theta)+\sqrt{(2\pi N_t-\theta)^2+1}\right|\right] \tag{6.5}$$

In the next sections, this arc-length $S$ from Eq. (6.5) is been employed for defining the cutting linearization error of the generated spiral tool trajectory.

### *6.1.2 Cutting Linearization Error*

In the following, the proposed surface analytical model (SAM) will be used to evaluate the cutting linearization error and to optimize the FTS/SSS diamond turning process. It will ensure that the tool trajectory in the cutting direction lies within the profile tolerance zone, as illustrated in Fig. 6.2.

Figure 6.3 demonstrates that a local $PV_{err}$ ($\delta_{err}$) between the targeted profile and the tool trajectory can be determined by identifying a maximum deviation $\partial Z_{\max}$

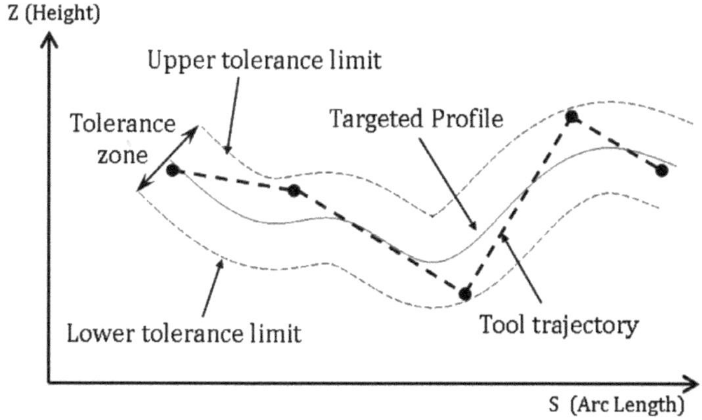


**Fig. 6.2** Surface generation of the tool trajectory should lie within the tolerance zone

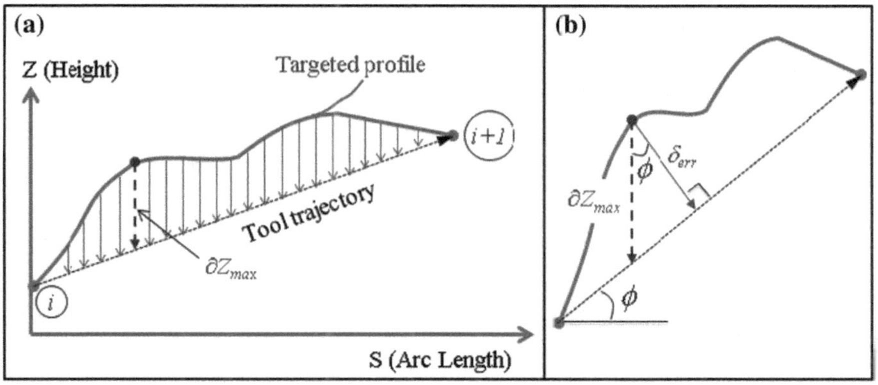


**Fig. 6.3** Schematic diagrams for determining $\delta_{err}$

between two corresponding cutting points. $\partial Z_{\max}$ represents the critical point on the ideal curve. $\delta_{err}$ is given as:

$$\delta_{err} = \partial Z_{\max} \cos \phi \tag{6.6}$$

where $\phi$ is the slope of tool trajectory.

During the process of evaluating the cutting linearization error, Newton's iteration method is employed in the surface analytical model to evaluate $\partial\theta$ with known $\partial S$ for determining the maximum deviation $\partial Z_{\max}$. From Eq. (6.4), the corresponding arc-length is

$$f(\theta) = \frac{f_r}{2}\left[\theta\sqrt{\theta^2+1} + \ln\left|\theta + \sqrt{\theta^2+1}\right| - S(\theta)\right] = 0, \tag{6.7}$$

Then, its derivative is

$$\begin{aligned} f'(\theta) &= \frac{f_r}{2}\left[\frac{2\theta^2+1}{\sqrt{\theta^2+1}} + \frac{1}{2}\frac{\frac{\theta+\sqrt{\theta^2+1}}{\sqrt{\theta^2+1}}}{\theta+\sqrt{\theta^2+1}}\right] \\ &= \frac{f_r}{2}\left[\frac{2\theta^2+1}{\sqrt{\theta^2+1}} + \frac{1/2}{\sqrt{\theta^2+1}}\right] \\ &= \frac{f_r}{2}\left[\frac{\theta^2+3/4}{\sqrt{\theta^2+1}}\right] \\ &\approx f_r\sqrt{\theta^2+1} \end{aligned} \tag{6.8}$$

Since the spiral tool trajectory begins from outermost radius, Eqs. (6.7) and (6.8) shall be rewritten as:

$$f(\partial\theta) = \frac{f_r}{2}\left[\begin{array}{c}(2\pi N_t - \partial\theta)\sqrt{(2\pi N_t - \partial\theta)^2 + 1} + \cdots \\ \ln\left|(2\pi N_t - \partial\theta) + \sqrt{(2\pi N_t - \partial\theta)^2 + 1}\right| - \partial S\end{array}\right] = 0 \tag{6.9}$$

for $\partial S \in [S_i, S_{i+1}]$ in a subtractive step $\leq 10^{-6}$ mm $(S_i > S_{i+1})$.

$$f'(\partial\theta) = f_r\sqrt{(2\pi N_t - \partial\theta)^2 + 1} \tag{6.10}$$

The iteration is continued until convergence has been obtained (i.e. $\partial\theta_j - \partial\theta_{j+1} \leq 10^{-9}$),

$$\partial\theta = \partial\theta - \frac{f(\partial\theta)}{f'(\partial\theta)} \tag{6.11}$$

From Eq. (3.1), the radial position $\partial\rho$ corresponded to $\partial\theta$ can be found as

$$\partial\rho = r - f_r\partial\theta \tag{6.12}$$

From Eq. (3.6), the ideal surface heights between two control points are

$$\partial Z_{ideal} = f(\partial\rho, \partial\theta) \tag{6.13}$$

The actual heights of the tool trajectory are

$$\partial Z_{tool} = Z_i + \partial S(Z_{i+1} - Z_i)/(S_i - S_{i+1}) \tag{6.14}$$

Finally, $\partial Z_{\max}$ is the maximum deviation between ideal surface height and actual height of the tool trajectory

$$\partial Z_{\max} = \arg\max(\partial Z_{ideal} - \partial Z_{tool}) \tag{6.15}$$

The entire procedure must be repeated for all corresponding control points to evaluate $PV_{err}$. The details for establishing the critical machining parameters will be explained in the case studies in the next section.

## 6.2 Experimental Validation

### *6.2.1 Evaluation of Critical Machining Parameters*

It is common knowledge that more control points yield a better profile accuracy. However, the maximum frequency of control point sequence should not be greater than the bandwidth of FTS/SSS system. Thus, a trade-off is necessary for achieving accurate profile with optimum number of control points.

**Table 6.1** Geometrical definitions for MLA surface

| Geometries | Values | Units |
|---|---|---|
| *Shape* (convex or concave) | Convex: −1 | – |
| Conic coefficient, $k$ | Sphere: 0 | – |
| Lens curvature, $C$ | 1/25 | $mm^{-1}$ |
| Lens radius, $r_{lens}$ | 0.5 | mm |
| Lens-lens pitch | 1.5 | mm |

Table reproduced from [6]

For evaluation, a microlens array (MLA) and a sinusoidal wave grid (SWG) given by

$$Z_{MLA} = \frac{shape \cdot Cr_{lens}^2}{1 + \sqrt{1 - (k+1)C^2 r_{lens}^2}} \tag{6.16}$$

$$Z_{SWG} = A_{SWG} \sin(2\pi x/\lambda) + A_{SWG} \cos(2\pi y/\lambda) \tag{6.17}$$

and will be fabricated. The parameters and their values are explained in Tables 6.1 and 6.2. A targeted profile accuracy tolerance ($PV_{tol}$) is 1.0 μm and the cutting feedrate $f_r$ is 0.0025/π mm/radian (or 0.005 mm/rev).

Figures 6.4 and 6.5 illustrate the 3D plots for the calculated $PV_{err}$ of the MLA and SWG surfaces, respectively, obtained with the two different cutting strategies. From Figs. 6.4a–d and 6.5a–d, the results demonstrate that the $PV_{err}$ on the outer radii regions under constant-angle cutting conditions are generally larger than those of inner regions. In contrast, the $PV_{err}$ results under constant-arc cutting conditions are almost uniform, as shown in Figs. 6.4e–h and 6.5e–h.

From Fig. 6.4d, g, the calculated results conclude that the critical values of the $\Delta\theta$ and $\Delta S$ parameters for the MLA surface are 0.5° and 0.0698 mm, respectively. and 2.0° and 0.0698 mm, respectively. On the other hand, the critical values of the $\Delta\theta$ and $\Delta S$ parameters for SWG surfaces, as shown in Fig. 6.5c, h, are 2.0° and 0.0698 mm, respectively. If those values are above the critical values, the $PV_{err}$ will be exceeded the required $PV_{tol}$ of 1.0 μm. Hence, the selection of the machining parameters, $\Delta\theta$ and $\Delta S$, must not be more than the critical values.

Notwithstanding the fact that the constant-arc method would give an almost constant surface accuracy, there is a noticeable 'sprue-shape' $PV_{err}$ found in the central region under all constant-arc cutting conditions as illustrated in Figs. 6.5 and 6.6. Although this type of $PV_{err}$ has never been reported in the literatures, it still plays a major role in optimizing overall surface accuracy under constant-arc

**Table 6.2** Geometrical definitions for SWG surface

| Geometries | Values | Units |
|---|---|---|
| Amplitude, $A_{SWG}$ | 0.00625 | mm |
| Wavelength, $\lambda$ | 3.0 | mm |

Table reproduced from [6]

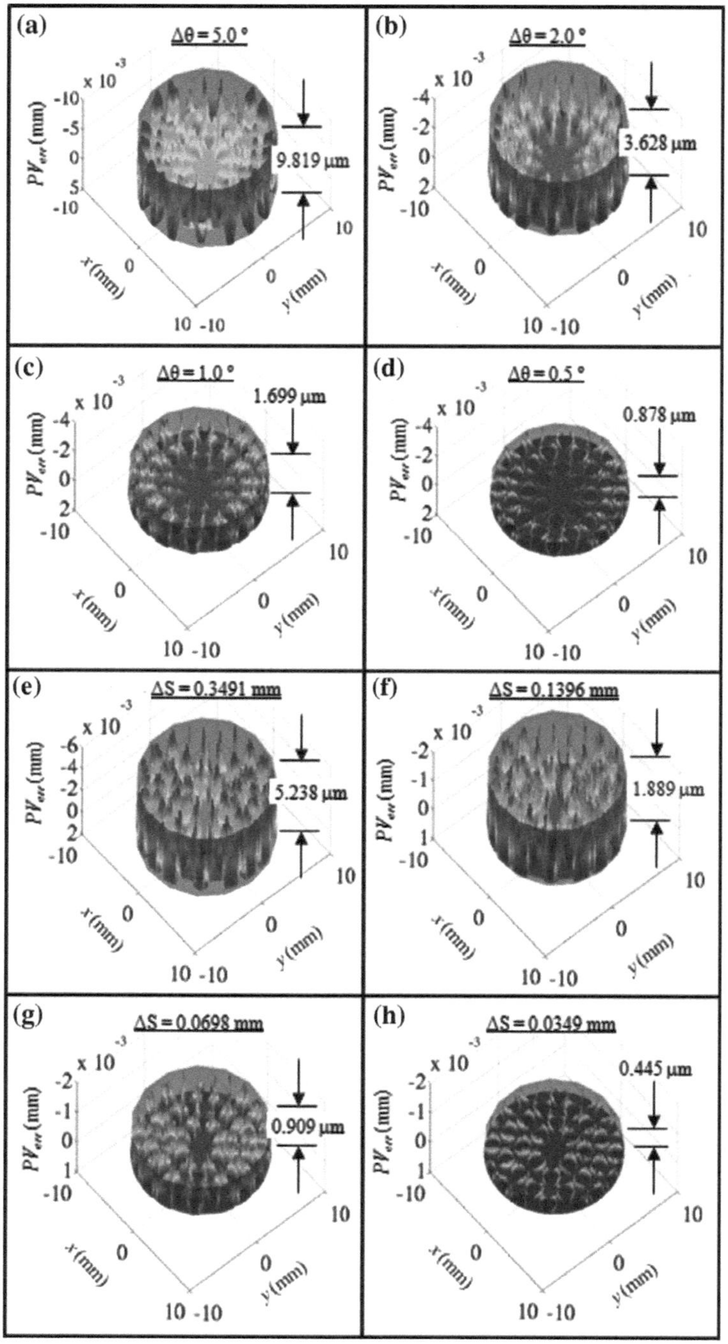


**Fig. 6.4** $PV_{err}$ results of the MLA surface under (**a–d**) $\Delta\theta$ and (**e**, **f**) $\Delta S$ cutting strategies. Figure reproduced from [6]

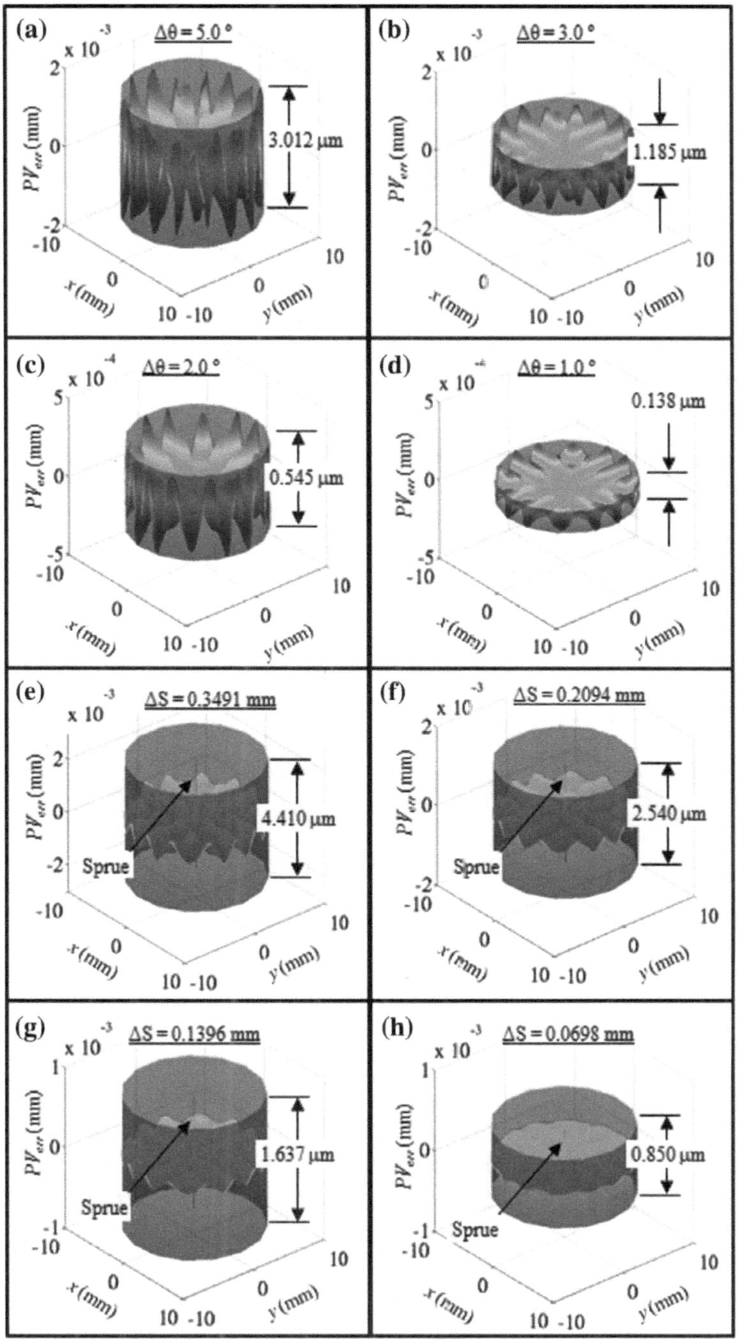


**Fig. 6.5** $PV_{err}$ results of the SWG surface under (**a**–**d**) $\Delta\theta$ and (**e**, **f**) $\Delta S$ cutting strategies. Figure reproduced from [6]

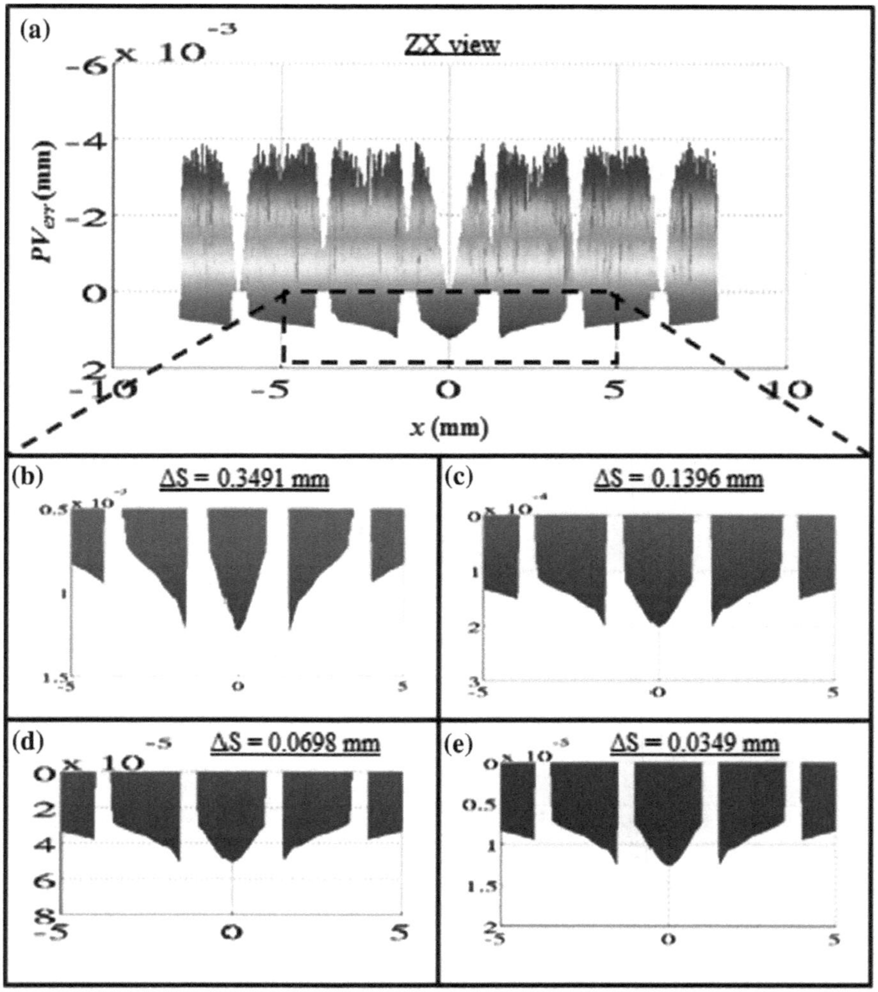


**Fig. 6.6** **a** $PV_{err}$ plot of the MLA surface (XZ view), and (**b–e**) the enlarged views of different constant-arc cutting conditions indicating the presence of 'sprue-shape' $PV_{err}$ in the central region. Figure reproduced from [6]

conditions. This can be explained by the fact that the arc-length $S$, as explained in Eq. (6.5) which is a function of its associated angle $\theta$. Thus, the associated angles in the central region are larger than those in the outer regions.

Furthermore, it also reveals that the $PV_{err}$ results in the outer regions are much lower than $PV_{tol}$ when the value of $\Delta S$ has been re-defined to reduce this 'sprue-shape' $PV_{err}$. On top of that, there will be an increase in the number of tool trajectory controlled points while reducing the value of $\Delta S$. As explained earlier, it is strongly recommended not to have a large number of controlled points as it would reduce the spindle speed degrading the machined surface quality. Thus, a

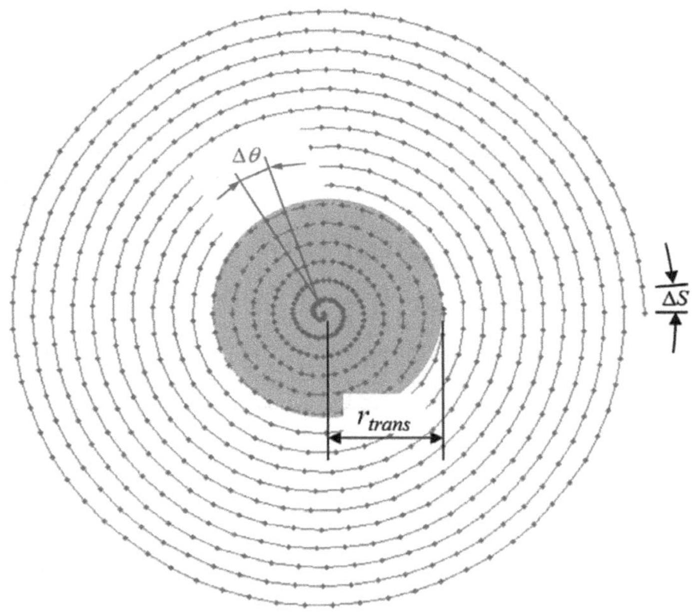


**Fig. 6.7** HCAA method of controlling tool trajectory

hybrid of constant-arc and constant-angle (HCAA) method has been employed for overcoming this 'sprue-shaped' $PV_{err}$ problem.

In the HCAA method, as illustrated in Fig. 6.7, the surface analytical model is first used to evaluate the critical $\Delta S$ at the outer radius of the surface using the constant-arc cutting strategy. Then, a transition radius $r_{trans}$ is identified as a radial location where the $PV_{err}$ exceeds the required $PV_{tol}$. After $r_{trans}$ is defined, the constant-angle method is applied from the point where the constant-arc method is abandoned to evaluate the critical $\Delta\theta$ for the inner region.

As can be seen from the analytical results shown in Fig. 6.8, the transition radiuses between outer and inner regions are located at 1.450 and 0.8195 mm for the MLA and SWG surfaces, respectively. The critical $\Delta S$ values for the outer regions of the MLA and SWG surfaces are 0.0698 and 0.3491 mm, respectively, and the critical $\Delta\theta$ for the central regions are 180° and 24°, respectively.

Finally, a comparative study between different cutting strategies has been conducted based on their optimum number of cutting points. Table 6.3 reveals that the HCAA method offers the optimum number of cutting points, especially only 109,252 cutting points for the SWG surface, which is about 20% of the number needed in conventional constant-arc method.

Furthermore, an interesting conclusion is that the constant-arc method is not necessarily superior to the constant-angle method, depending on the type of free-form surfaces to be machined. For example, the constant-arc method yields a smaller number of cutting points than the constant-angle method for the MLA

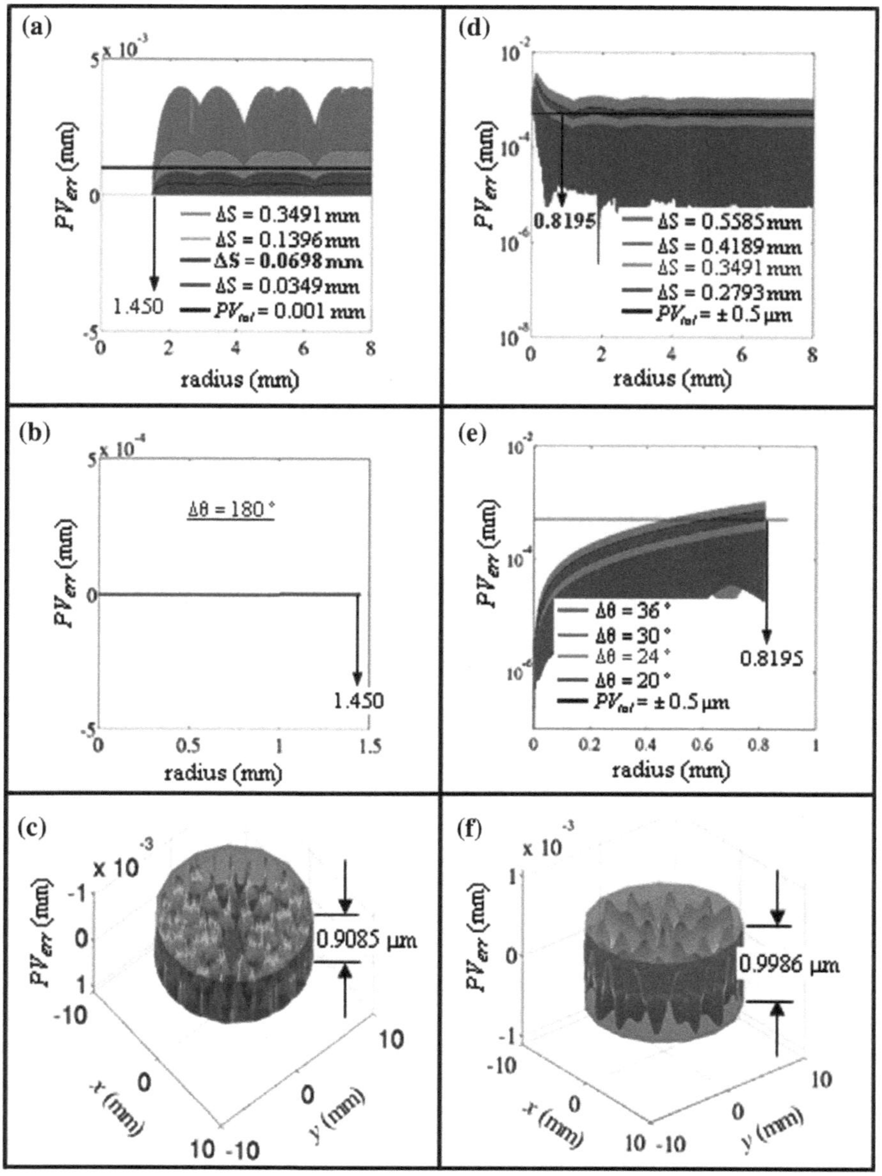


**Fig. 6.8** Application of the HCAA method of evaluating the critical parameters for (**a–c**) MLA and (**d–f**) SWG surfaces

surface, but more cutting points than constant-angle method for the SWG surface. Thus, the selection of the cutting strategy should also be optimized for ultraprecision machining of freeform surfaces.

**Table 6.3** Comparison between different cutting strategies

| Type of surfaces | Optimum number of trajectory points under different cutting strategies | | | Percentage reduction | |
|---|---|---|---|---|---|
| | $\Delta\theta$ | $\Delta S$ | HCAA | HCAA versus $\Delta\theta$ (%) | HCAA versus $\Delta S$ (%) |
| MLA | 1,080,001 | 540,001 | 521,950 | 51.67 | 3.34 |
| SWG | 270,001 | 540,001 | 109,252 | 59.53 | 79.77 |

## 6.2.2 Cutting Experiments and Results

All the experiments were conducted by a slow slide servo (SSS) diamond turning process. Based on the novel surface analytical model as described in the previous sections, the critical cutting parameters for different cutting strategies were optimized and selected in machining SWG and MLA surfaces to meet the required PV tolerance of 1.0 μm, as shown in Table 6.4.

These machined surfaces were measured using an Olympus LEXT OLS4000 3D Measuring Laser Microscope and the measured 3D profile data were further post-processed for surface characterization. For evaluation of the profile errors of the fabricated features, areal error maps were used along with the characterization method by Yu et al. [7]. First, the measured data were extracted as CSV files to be processed using MATLAB software. Second, a best fit of the measured profile was performed for a comparison with the theoretical surface using a rigid body transformation described the matrix [7]:

$$T(t_x, t_y, t_z, \alpha, \beta, \gamma) = \begin{bmatrix} \cos\beta\cos\gamma & \sin\alpha\sin\beta\cos\gamma - \cos\alpha\sin\gamma & \cos\alpha\sin\beta\sin\gamma + \sin\alpha\sin\gamma & t_x \\ \cos\beta\sin\gamma & \sin\alpha\sin\beta\sin\gamma + \cos\alpha\cos\gamma & \cos\alpha\sin\beta\sin\gamma - \sin\alpha\cos\gamma & t_y \\ -\sin\beta & \sin\alpha\cos\beta & \cos\alpha\cos\beta & t_z \\ 0 & 0 & 0 & 1 \end{bmatrix} \tag{6.18}$$

**Table 6.4** Selected cutting conditions for machining MLA and SWG surfaces

| Parameters | Values |
|---|---|
| Spindle speed | 50 rev/min |
| Feedrate, $f_r$ | 0.005 mm/rev (rough) and 0.001 mm/rev (finish) |
| Depth of cuts | 0.010 mm (rough) and 0.005 mm (finish) |
| *Cutting strategies* | |
| (i) Constant-angle | $\Delta\theta = 0.5°$ (MLA)<br>$\Delta\theta = 2.0°$ (SWG) |
| (ii) Constant-arc | $\Delta S = 0.0698$ mm (for both MLA and SWG) |
| (iii) HCAA | |
| For outer region | $\Delta S = 0.0698$ mm (MLA) and 0.3491 mm (SWG) |
| For inner region | $\Delta\theta = 180°$ (for both MLA and SWG) |

where $t_x$, $t_y$ and $t_z$ are translations in the $x$, $y$ and $z$-directions respectively, and $\alpha$, $\beta$, $\gamma$ are the rotations about $x$, $y$ and $z$-directions, respectively.

Figure 6.9 illustrates the photographic images of machined SWG surfaces with their respective measured 3D profiles and contour errors under three different cutting strategies, namely constant-angle, constant-arc and HCAA. It can be seen that the evaluated results of $PV_{err}$ in the $x$- and $y$-directions for all cutting strategies

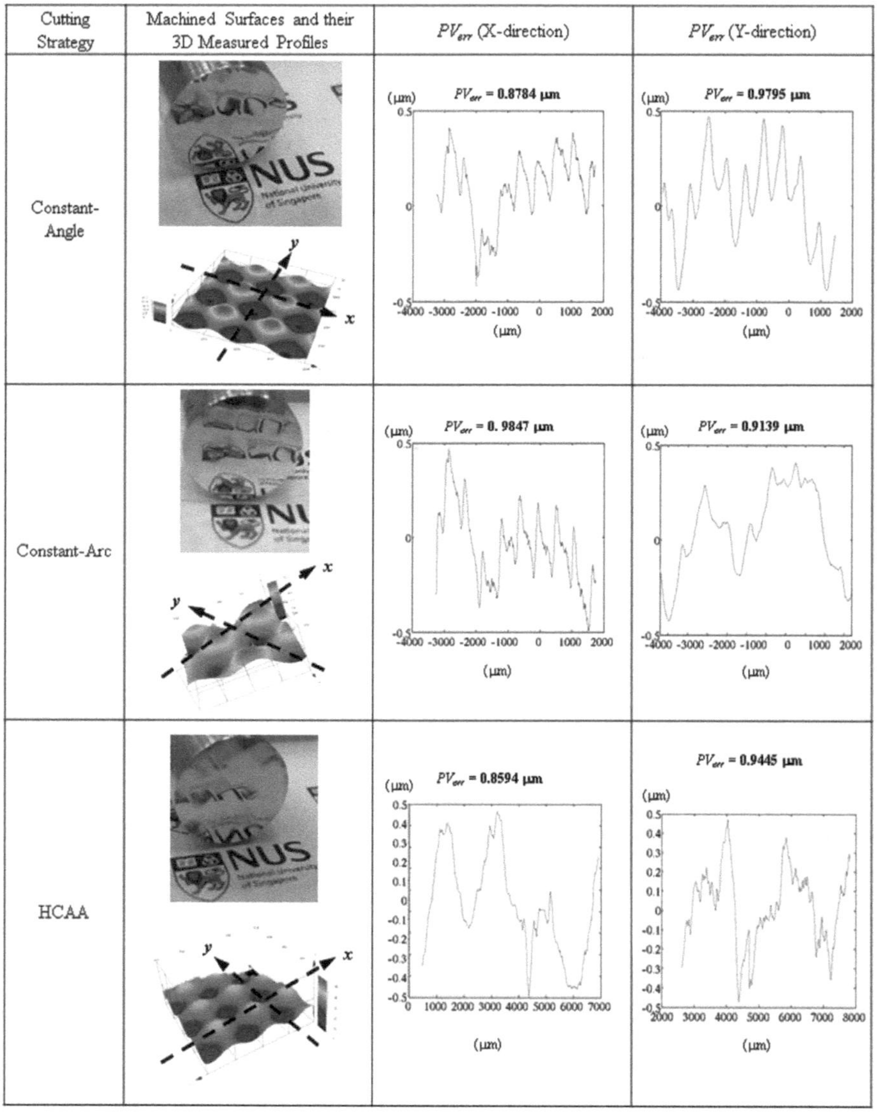


**Fig. 6.9** Contour error of the SWG surface under (*top*) $\Delta\theta$, (*middle*) $\Delta S$, and (*bottom*) HCAA cutting strategies

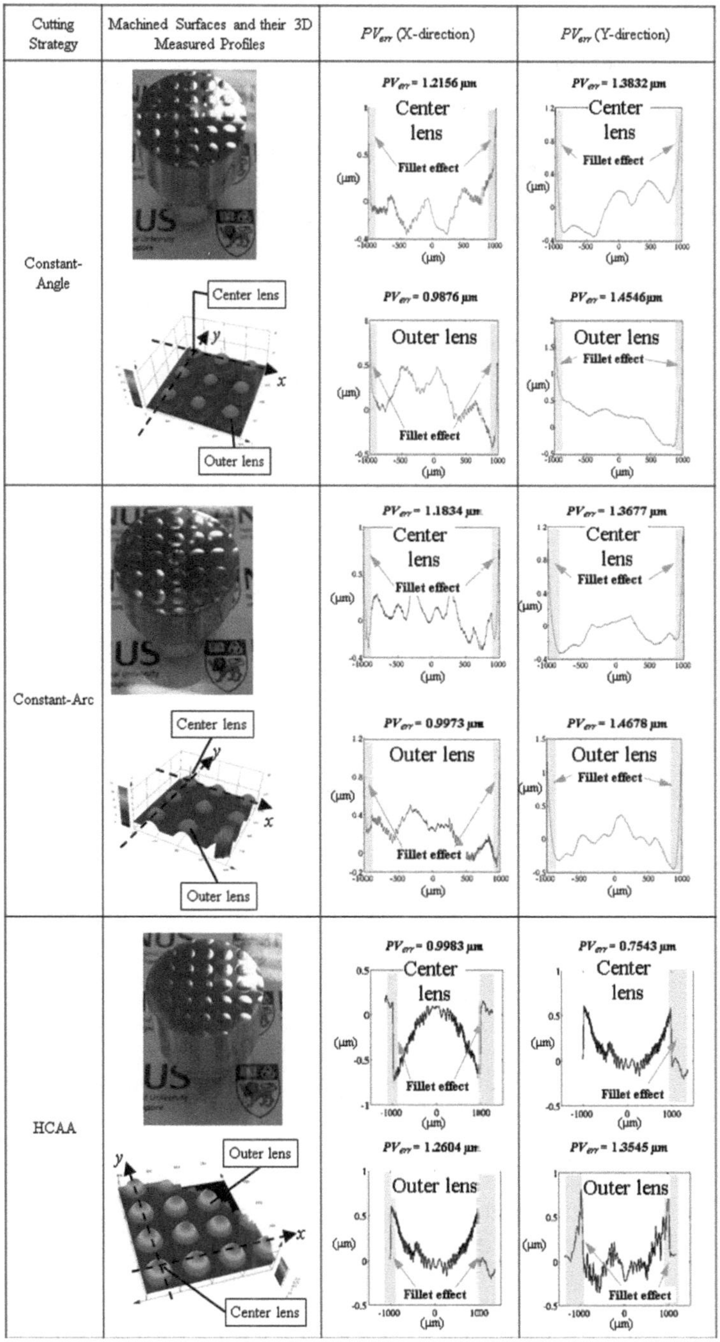


**Fig. 6.10** Contour error of the MLA surface under (*top*) $\Delta\theta$, (*middle*) $\Delta S$, and (*bottom*) HCAA cutting strategies

are validating the $PV_{tol}$ of 1.0 μm. These results also further validate that the selected cutting parameters (as in Table 6.4) for machining SWG surface using all different cutting strategies are properly defined.

Figure 6.10 illustrates the photographic images of machined MLA surfaces with their respective measured 3D profiles and contour errors under three different cutting strategies, namely constant-angle, constant-arc and HCAA, respectively. Fillet errors due to the tool nose radius are detected at the edges of the lenses, which cause the $PV_{err}$ failing to meet $PV_{tol}$ of 1.0 μm. Otherwise, the overall $PV_{err}$ results of the MLA surface with the selected parameters would have met the $PV_{tol}$ if neglected the fillet effect.

Lastly, these experimental results validate the credibility of the proposed surface analytical model for determining the optimal machining conditions for all cutting strategies. The model not only improves the profile accuracy, but also eliminates the risk of the machined workpiece to miss the accuracy requirements. Furthermore, it is concluded that the constant-arc cutting strategy is not necessarily be superior than the constant-angle method, depending on the curvatures of freeform surfaces.

## 6.3 Concluding Remarks

In the present study, a surface analytical model has been developed to evaluate the cutting linearization error on freeform surfaces with an optimal number of cutting points for the constant-angle ($\Delta\theta$), constant-arc ($\Delta S$) and hybrid (HCAA) methods. A good performance of the fast tool servo or slow slide servo alone is insufficient for producing accurate surfaces; it is also necessary to optimize the tool trajectory. The model developed in this study was experimentally verified by machining of a sinusoidal wave grid (SWG) and a micro-lens array (MLA) and evaluating the contour errors.

The profile accuracy requirement for all surfaces could be met. The surface evaluation has demonstrated that the Peak-to-Valley errors ($PV_{err}$) were within the specified tolerances. Moreover, the HCAA method combines the merits of both the constant-arc and constant-angle method for producing highly accuracy surfaces with an optimum number of control points. Offering a significant reduction of machining time and increased profile accuracies, the hybrid method is proposed as the new standard in surface generation.

The developed surface analytical model is later implemented and integrated with the developed CAD/CAM system in the SolidWorks-API environment to generate accurate ultraprecision hybrid freeform surfaces. The details of the integration are presented and discussed in the next chapter.

# References

1. Zhou M, Zhang HJ, Chen SJ. Study on diamond cutting of nonrationally symmetric microstructured surfaces with fast tool servo. Mat Manu Proc. 2010;25(6):488–94.
2. Liu K, Wu H, Liu P, Shaw KC. Ultra-precision machining of aluminium alloy surfaces for optical application. Int J Nanomanuf. 2011;7(2):116–25.
3. Yu DP, Wong YS, Hong GS. Optimal selection of machining parameters for fast tool servo diamond turning. Int J Adv Manuf Technol. 2011;57:85–99.
4. Weisstein EW. Arc length, from MathWorld—a Wolfram web resource. http://mathworld.wolfram.com/ArcLength.html.
5. Polyanin AD, Manzhirov AV. Supplement 3. Tables of indefinite integrals in handbook of integral equations. 2nd ed. Chapman & Hall/CRC, Taylor & Francis Group; 2008. p. 933–50.
6. Neo DWK, Kumar AS, Rahman M. A novel surface analytical model for cutting linearization error in fast tool/slow slide servo diamond turning. Precis Eng. 2014;38(4):849–60.
7. Yu DP, Zhong X, Wong YS, Hong GS, Lu WF, Cheng HL. An automatic form error evaluation method for characterizing micro-structured surfaces. Meas Sci Technol. 2011;22 (1):015105. doi:10.1088/0957-0233/22/1/015105.

# Chapter 7
# Integration and Implementation

This chapter presents an overall integration of the proposed methodologies throughout this dissertation. The primary objective of the integrated system is to construct the software tools for planning and conducting the manufacture of hybrid freeform surfaces using the multiple-axis ultraprecision machining process.

Section 7.1 presents the overall view of the developed integrated system and controller configuration, and also discusses the optimization of tool geometry in the integrated system for avoiding the tool interference onto the machined surface. Sections 7.2 and 7.3 present the two case studies to validate the creditability of the developed integrated system. Lastly, the concluding remarks are discussed in Sect. 7.4.

## 7.1 Integrated CAD/CAM System

The screenshot image of the main menu for the developed SolidWorks-API user-interface in the integrated CAD/CAM system is shown in Fig. 7.1. It can be seen that there are two different cutting processes, namely automated Guilloche machining technique (AGMT) and diamond turning processes. The user selects one of the cutting processes for the manufacturing of hybrid freeform surfaces. The details of two developed sub-systems and the configuration requirements of the controllers are discussed in the next sections. In addition, a tool optimization process and geometrical splitting of hybrid freeform surface also embedded into the system and its details are discussed in Sects. 7.1.3 and 7.1.4, respectively.



D.W.K. Neo, *Ultraprecision Machining of Hybrid Freeform Surfaces Using Multiple-Axis Diamond Turning*, Springer Theses,
DOI 10.1007/978-981-10-4083-2_7

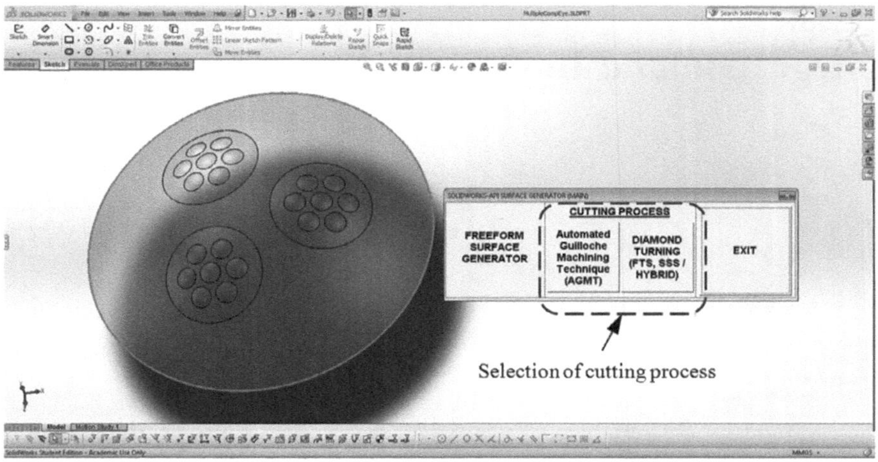


**Fig. 7.1** Screenshot image for main menu of user-interface in the integrated system for the selection of a cutting process. Some content of this chapter has been reproduced with permission from [3, 9]

## 7.1.1 Integrated Sub-system for AGMT Process

The screenshot image of the developed user-interface for the AGMT process in a sub-system of the integrated system is illustrated in Fig. 7.2. First of all, the user has to select the type of Fresnel lens arrays to be machined and its layout. Then, the user shall input the lens dimensions, namely radius/apothem (depending on types of Fresnel lens) and pitch distance, before selecting the surfaces to be machined. Next, the user has to define all machining and tool parameters, and targeted accuracy errors for evaluating the critical machining parameters and profile errors by the developed integrated system. These critical parameters would generate at the end of evaluation process and display in the user-interface form. Then, the user would decide to keep or modify these critical values before post-processing the cutting points as NC-codes for the fabrication of Fresnel lens array.

Furthermore, there is another option for exporting the detailed descriptions of the calculated critical parameters into a data file for the user to understand the results clearly. The developed sub-system for AGMT process would be validated with a case study. The details are presented and discussed in Sect. 7.2.

## 7.1.2 Integrated Sub-system for Diamond Turning Process

The screenshot image of the developed user-interface for diamond turning process in another sub-system of the integrated system is shown in Fig. 7.3. Similarly, the user has to select a type of diamond turning processes and its cutting strategy,

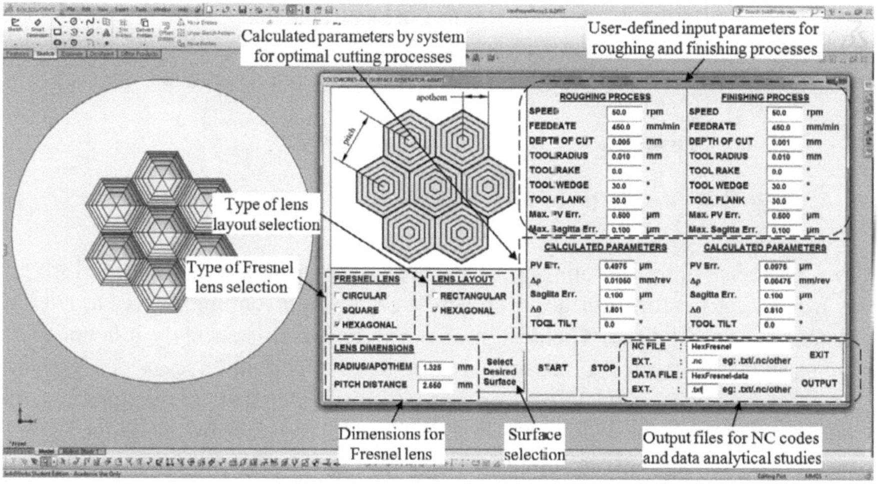


**Fig. 7.2** Screenshot image for user-interface of sub-system to generate the Guilloche tool trajectory for AGMT process

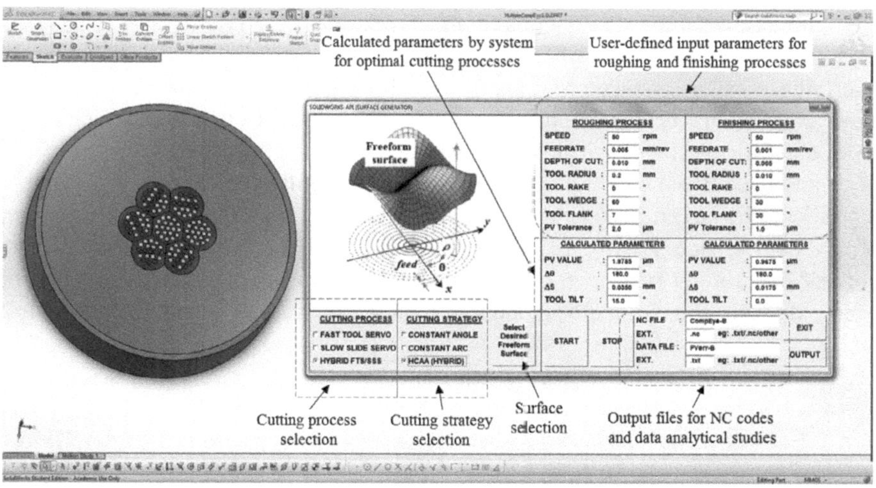


**Fig. 7.3** User interface of SolidWorks-API to generate the spiral tool trajectory for hybrid FTS/SSS process. Figure reproduced from [3]

before selecting the surfaces to be machined. Next, the user has to define all machining and tool parameters, and targeted accuracy errors for evaluating the critical machining parameters and profile errors by the developed integrated system. These critical parameters would generate at the end of evaluation process and display in the user-interface form. Then, the user would decide to keep or modify these critical values before post-processing the cutting points as NC-codes for the fabrication of freeform surface. Lastly, the user may export the detailed descriptions of the calculated critical parameters into a data file for the analytical studies. The developed sub-system for diamond turning process would be validated with a case study. The details are presented and discussed in the later sections.

### 7.1.3 Optimization of Tool Geometry

A tool geometry optimization process has been developed and implemented for not only machining freeform surfaces efficiently, but also preventing the tool to overcut or damage the workpiece. The design of a diamond tool is strongly influenced by the curvatures of freeform surface. Hence, it is important to consider the tool geometries, namely rake ($\gamma$), clearance ($\alpha$) and included ($\beta$) angles in the selection/designing a new diamond tool. Although there are few developed mathematical models [1, 2] to optimize the tool geometry, they are limited to two-dimensional surface curvatures along the cutting and feed directions but not for three-dimensional surface curvatures. Thus, the methodologies have been proposed to come the above limitations by obtaining these critical tool geometrical angles directly from CAD system.

Figure 7.4 illustrates that there are three types of cutting interference to be taken into an account when selecting or designing a new diamond tool. Firstly, a rake interference occurs between tool rake face and the surface curvature when the tool rake$\gamma_{tool}$ is highly negative. This can be overcome by selecting $\gamma_{tool}$ which is not more the critical rake $\gamma_{cr}$. Secondly, a flank interference occurs between tool flank faces and the surface curvature due to insufficient tool flank clearance $\alpha_{tool}$. Hence, $\alpha_{tool}$ should be larger than critical front clearance $\alpha_{cr}$ so that the flank faces are free of any interference. Lastly, a side interference is also necessary to be avoided to prevent overcutting of side cutting edges onto the surface curvature. This can be overcome when the tool included angle $\beta_{tool}$ is lesser than the critical included angle

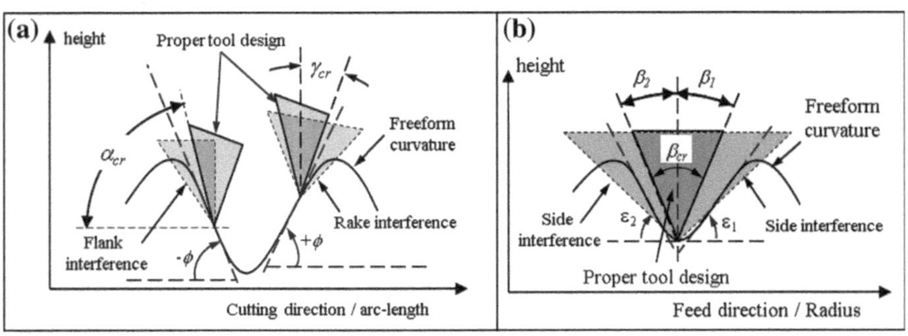


**Fig. 7.4** Types of cutting interference and critical tool angles

$\beta_{cr}$. These critical tool geometrical angles can be defined with the following conditions:

(i) cutting direction's boundary conditions:

$$\begin{array}{ll} \gamma_i = \pi - \phi_i - C, & \text{if } \phi_i \geq 0 \\ \alpha_i = |\phi_i| + C, & \text{if } \phi_i < 0 \\ \gamma_{cr} \leq \arg\max(\gamma_i) & \\ \alpha_{cr} \geq \arg\max(\alpha_i) & \end{array} \tag{7.1}$$

(ii) feed direction's boundary conditions:

$$\begin{array}{ll} \beta_1 = 90^\circ - \varepsilon_i - C, & \text{if } \varepsilon_i \geq 0 \\ \beta_2 = 90^\circ - |\varepsilon_i| - C, & \text{if } \varepsilon_i < 0 \\ \beta_i = \beta_1 + \beta_2 & \\ \beta_{cr} < \arg\max(\beta_i) & \end{array} \tag{7.2}$$

where $\phi$ and $\varepsilon$ are the slopes of the freeform curvature along the cutting and feed directions, respectively, and $C$ is a clearance angle to prevent tool faces overcutting/interfering against the machined surface.

Lastly, the immediate availability of standard tool sizes is also an important factor in designing these critical angles, especially $\alpha_{cr}$. Although a diamond tool is custom-made with respect to the critical tool geometrical angles, it would be not only costly but also required a long lead time. Furthermore, the maximum $\alpha_{tool}$ for the manufacturers to produce is 30° or lesser and may not be able to meet the requirement of $\alpha_{cr}$. Hence, these drive the needs for alternative solution of designing diamond tool geometry at a shortest time with an economical cost [3]. The front clearance issues can be overcome by titling the tool holder, as illustrated in Fig. 7.5, and both $\alpha$ and $\gamma$ would be also adjusted accordingly. Thus, the critical tilted angle $\lambda_{cr}$ can be defined as:

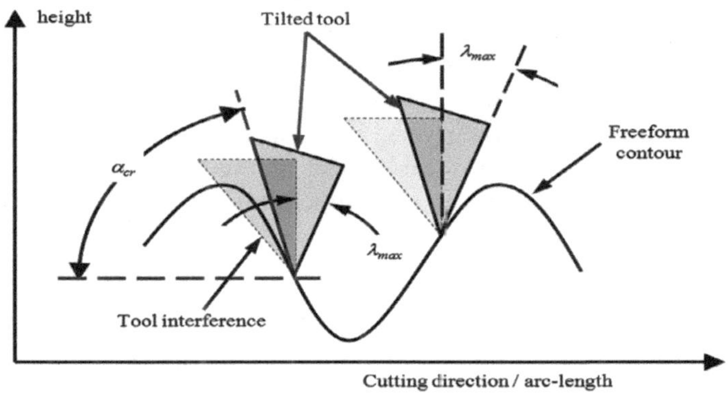


**Fig. 7.5** Schematic diagram for titling a tool holder

$$\lambda_{cr} = \alpha_{cr} - \alpha_{tool} \tag{7.3}$$

The details for determining the critical tool geometrical angles will be elaborated with case studies in the later sections.

### 7.1.4 Geometrical Splitting of Hybrid Freeform Surface

A hybrid freeform surface may come in a hybrid form of several freeform surfaces, which is possible to manufacture by employing a hybrid FTS/SSS diamond turning process. Although the layered tool trajectory methodology has been developed in the previous studies (Chap. 4), this methodology is only designed for micro prism but not for any freeform feature. Hence, it is necessary to have a proper approach to split up the hybrid freeform surface into two types of features, which would be machined by FTS and SSS processes, simultaneously. In this study, the splitting up process can be done by based on their geometrical frequency properties using the fast Fourier transform (FFT) and Hilbert transform (HT) approaches.

FFT defines the generated tool trajectories into a different domain with several distinguished properties for detailed analysis [4], and the discrete Fourier transform of the generated tool trajectory signals *dZ* is given by:

$$\begin{aligned} dZ(k) &= \sum_{j=1}^{N} \partial z(j)\omega_N^{(j-1)(k-1)} \\ \omega_N &= e^{\left(-2\pi\sqrt{-1}\right)/N} \end{aligned} \tag{7.4}$$

where $\partial z$ is the incremental adjustment of $z$-values between two corresponding tool trajectory points.

Due to the fact that a tool trajectory from HCAA cutting strategy has multiple frequencies, it would be difficult to implement the FFT approach. Thus, Hilbert transform approach is necessary to overcome this multiple frequencies problem. HT approach has been widely employed as an important tool in different branches of science and technology, from complex analysis and optics, to circuit theory and control science [5, 6]. HT not only behaves similarly to FFT but also is a linear operator, making it useful for analyzing non-stationary signals by expressing frequency as a rate of change in phase, so that the frequency can vary with time. Hence, HT is often introduced as a convolution between $f(x)$ and $-1/(\pi x)$ [7]:

$$H(f(x)) = -\frac{1}{\pi x} * f(x) = -\frac{1}{\pi}\int_{-\infty}^{\infty} f(x')\frac{dx'}{x' - x} \tag{7.5}$$

If $x(n)$ is a causal and absolutely summable real sequence with a discrete time Fourier transform $X(e^{j\omega})$, then Hilbert transform can rewritten [8]:

$$X_{\text{im}}\left(e^{j\omega}\right) = -\frac{1}{2\pi}\int_{-\pi}^{\pi} X_{re}\left(e^{j\omega}\right)\coth\left(\frac{\omega-\varphi}{2}\right)d\varphi$$
$$X_{re}\left(e^{j\omega}\right) = x(0) + \frac{1}{2\pi}\int_{-\pi}^{\pi} X_{\text{im}}\left(e^{j\omega}\right)\coth\left(\frac{\omega-\varphi}{2}\right)d\varphi \tag{7.6}$$

where $X_{re}(e^{j\omega})$ and $X_{re}(e^{j\omega})$ are the real and imaginary parts of $X(e^{j\omega})$.

The details for determining the geometrical splitting of hybrid freeform surface based on their discrete time frequencies will be elaborated with a case study in the later sections.

## 7.2 Case Study 1: Hexagonal Fresnel Lens Array Using AGMT Process

This case study presents the machining of hexagonal Fresnel array using AGMT process [9]. Fresnel lens array is one of these hybrid freeform surfaces, which may have facet groove orientations arranged in a rectangular or hexagonal layout to form a polygonal Fresnel lens [10–12]. Notwithstanding the fact that a radius of Fresnel profile in such layouts could be the same as that of circular one (Fig. 7.6), the facet grooves are parallel to an apothem. It is possible to manufacture such arrangements with extrusion process and followed by assembly process to form together within the congruent triangles having a common vertex at the centre of the polygon. However, this approach leads to the misalignment of assembled profiles which may impact the overall optical performance. Hence, this misalignment issue could be eliminated by employing the developed automated Guilloche machining technique (AGMT) with a linear tool motion, as demonstrated in Fig. 7.7.

### *7.2.1 Experimental Validations*

In this case study, the developed automated Guilloche machining technique was employed for machining an array of hexagonal Fresnel lenses, as illustrated in Fig. 7.8. Table 7.1 describes the machining parameters which are the inputs for the developed user-interface of the sub-system. The surface generation for the Guilloche tool trajectory was conducted by implementing the methodologies into the developed integrated system.

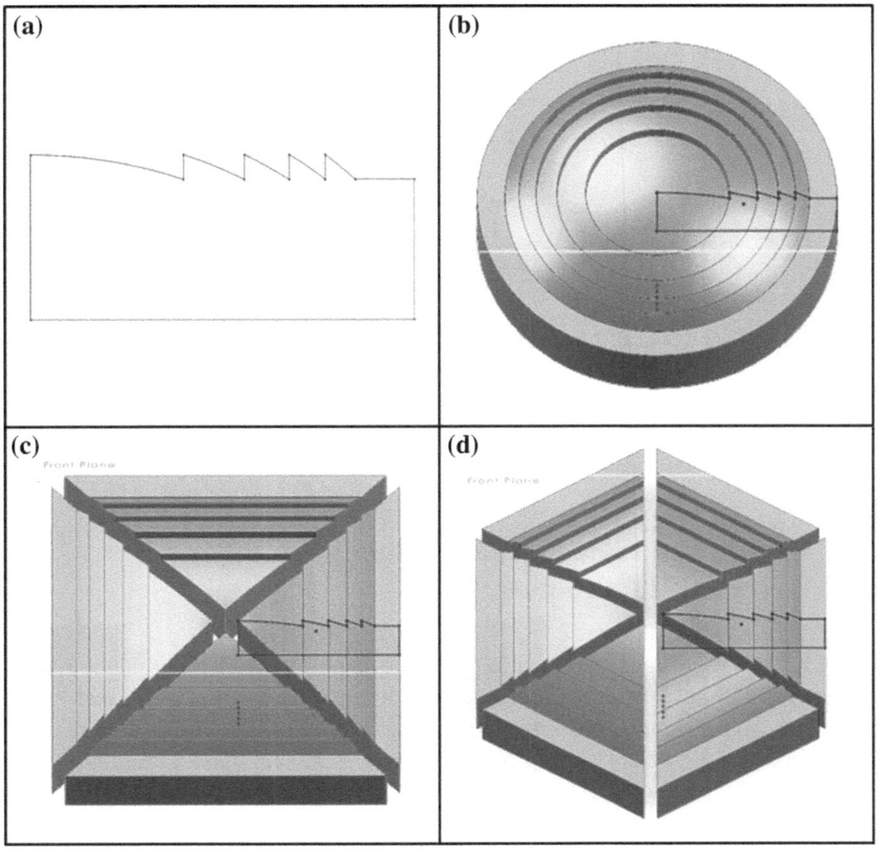


**Fig. 7.6** Fresnel lens designs; **a** cross-sectional profile of Fresnel zone, **b**-**d** circular, square and hexagonal types, respectively

Figure 7.9a and b show that the cutting points were successfully mapped onto the surface of central and offset hexagonal Fresnel lenses, respectively. These cutting points has been post-processed into the Guilloche tool trajectories, as illustrated in Fig. 7.9c and d, for the respectively lenses.

At the same time, the critical machining parameters for the generated cutting points have also been pre-evaluated by the developed integrated system, as demonstrated in Fig. 7.10, before the tool trajectory points are further post-processed into NC-codes. It can be seen that the critical feedrates $\Delta\rho_{cr}$ for roughing and finishing processes are 0.01050 and 0.00475 mm, respectively. The critical angular pitching for roughing and finishing processes $\Delta t_{cr}$ are 1.801° and 0.810°, respectively. Furthermore, the detailed information of these critical data are replicated from the developed API system and discussed in the next sections.

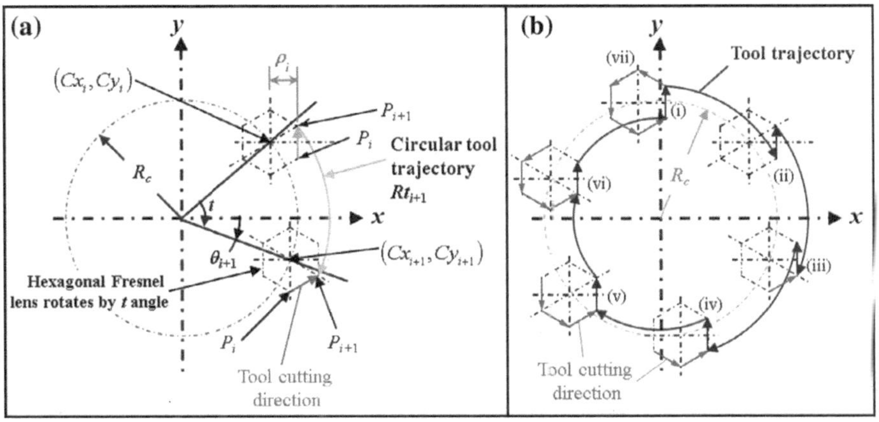


**Fig. 7.7** Schematic diagram of machining hexagonal Fresnel lens array; **a** calculation of tool control points, **b** Guilloche tool trajectory for one rotation from Steps (*i*) to (*vii*)

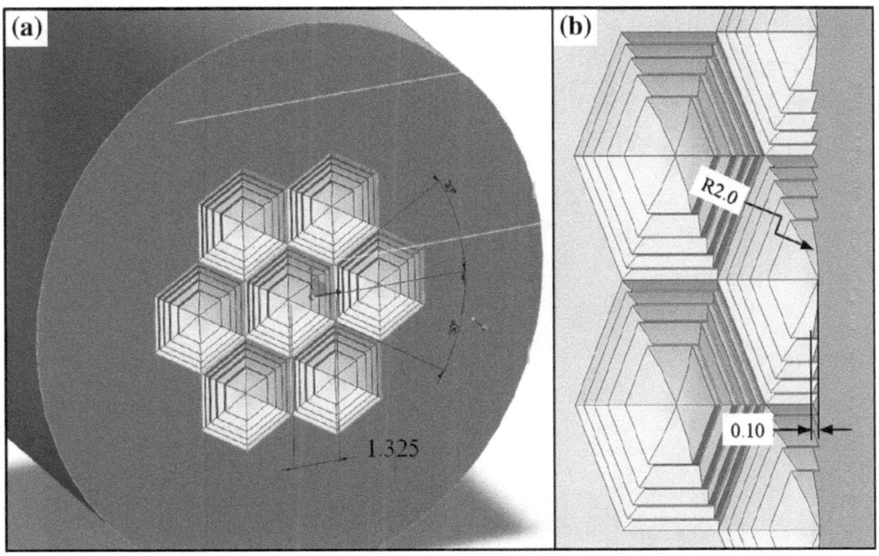


**Fig. 7.8** A CAD model for hexagonal Fresnel lens array; **a** in slanted view, and **b** half-sectional view

#### 7.2.1.1 Critical Machining Parameters for AGMT Process

Figure 7.11 shows the critical results of feedrate $\Delta\rho_{cr}$ for roughing and finishing processes, which are reproduced from the developed system. The selection of feedrates for roughing and finishing processes should not be more than $\Delta\rho_{cr}$ of

**Table 7.1** Input parameters for fabrication of hexagonal Fresnel lens array

| Machining parameters | Value | Units |
|---|---|---|
| Spindle speed | 50.0 | rpm |
| Linear feedrate | 450 | mm/min |
| Workpiece | AA6061-T6 | – |
| Tool | Mono-crystalline diamond, 10 μm nose radius, 40° front clearance | – |
| Max. cutting residual error, $E_\rho$ | 0.500 (roughing cuts)<br>0.100 (finishing cuts) | μm |
| Max. Sagitta error, $h_{tol}$ | 0.500 (roughing cuts)<br>0.100 (finishing cuts) | μm |

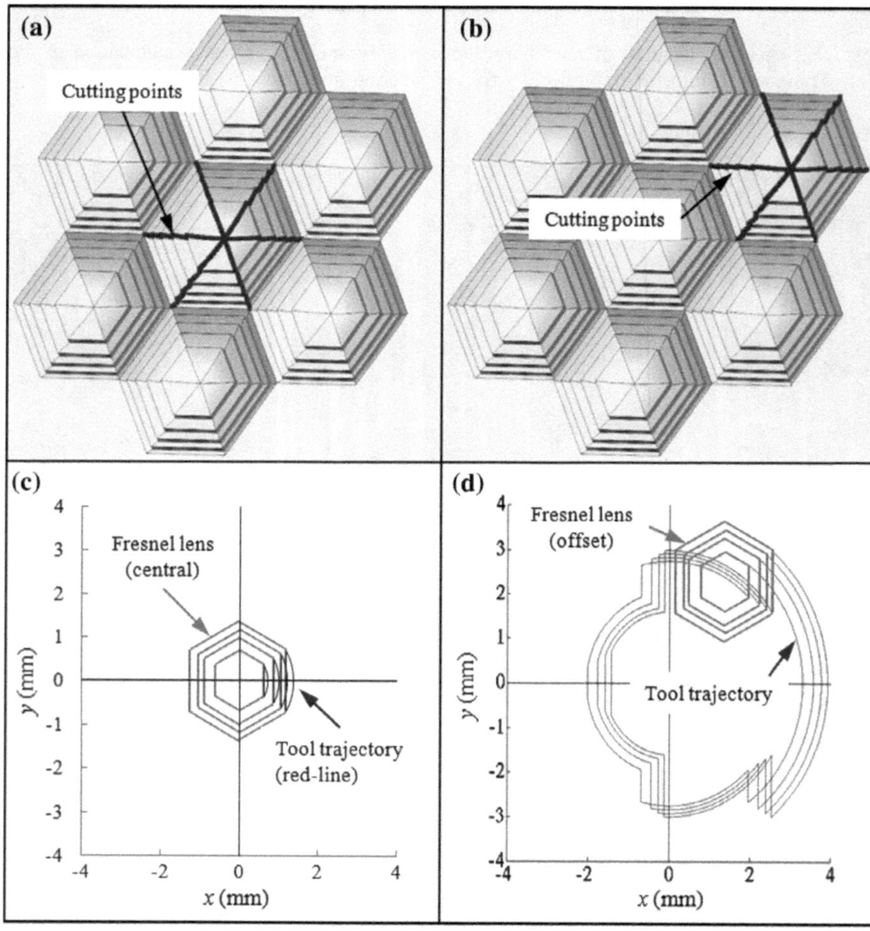


**Fig. 7.9** Successful generation of Guilloche tool trajectory points using the developed integrated system; Cutting points were mapped on central and offset Fresnel lenses in (**a–b**) respectively, and (**c–d**) the simulated Guilloche tool trajectories were conducted successfully on the respectively lenses

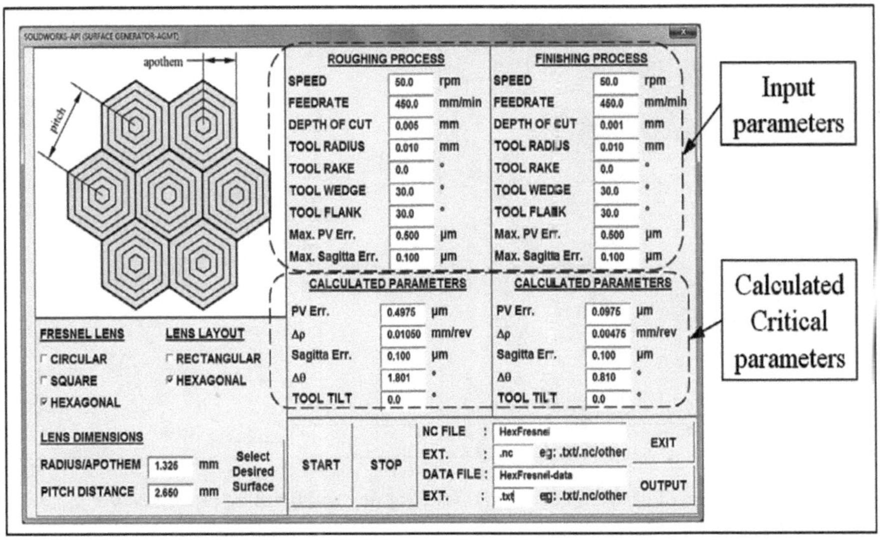


**Fig. 7.10** Screenshot image for the output results of calculated critical parameters for optimal AGMT process by the developed integrated system

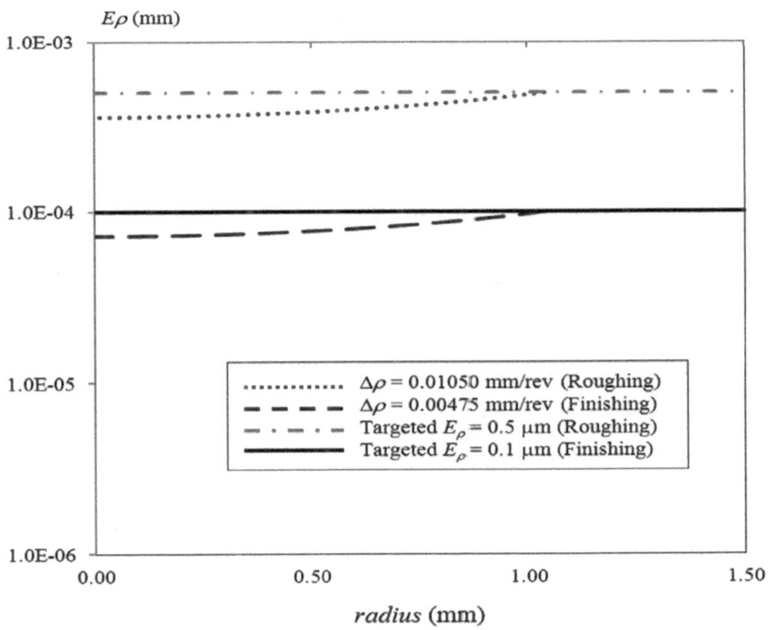


**Fig. 7.11** Critical results of roughing and finishing feedrates with given $E_{\rho}$, reproduced from output data from developed system

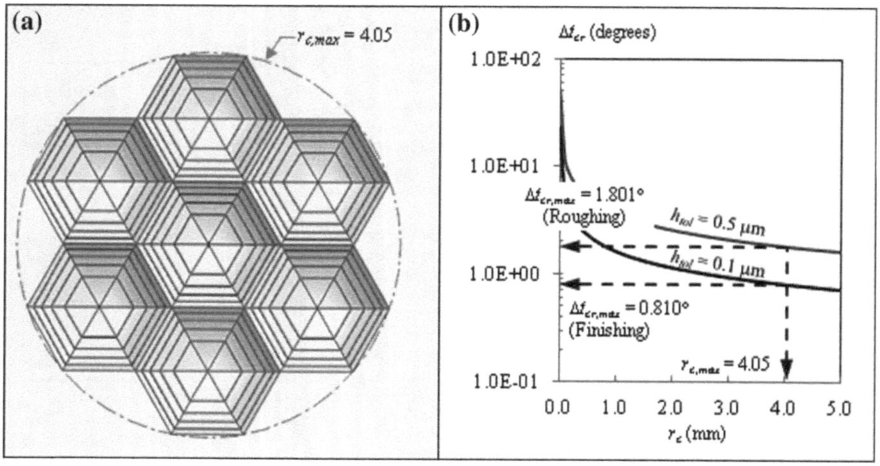


**Fig. 7.12** Calculation of critical pitch angular for roughing and finishing processes; **a** $r_{c,\ \max}$ of an array, and **b** critical $\Delta t_{cr}$

0.01050 and 0.00475 mm, respectively. Otherwise, it would not be able to meet the requirements of $E_\rho$ (as given in previous Table 7.1).

On the other hand as illustrated from Fig. 7.12, the reproduced critical values of angular pitching $\Delta t_{cr}$ based on a given $r_{c,\ max}$ of 4.05 mm are 1.801° and 0.810° for roughing and finishing processes, respectively. Hence, the selection of angular pitching should not be more than $\Delta t_{cr}$ in order to fulfill the requirement of targeted $h_{tol}$ (refer to Table 7.1).

#### 7.2.1.2 Cutting Experiments and Results

The automated Guilloche machining technique, AGMT [9], has been employed to perform the cutting experiments. Table 7.2 describes the cutting conditions have been selected based the above approaches to meet the targeted requirements. Figure 7.13 illustrates the photographic images for the successful fabrication of an array of hexagonal Fresnel lenses. These machined Fresnel lenses were measured

**Table 7.2** Selected cutting conditions for hexagonal Fresnel lens array

| Machining parameters | Value | Units |
|---|---|---|
| Feed, $\Delta\rho$ | 0.010 (roughing cuts)<br>0.0025 (finishing cuts) | mm/rev |
| Angular pitching, $\Delta t$ | 1.50 (roughing cuts)<br>0.75 (finishing cuts) | (°) |

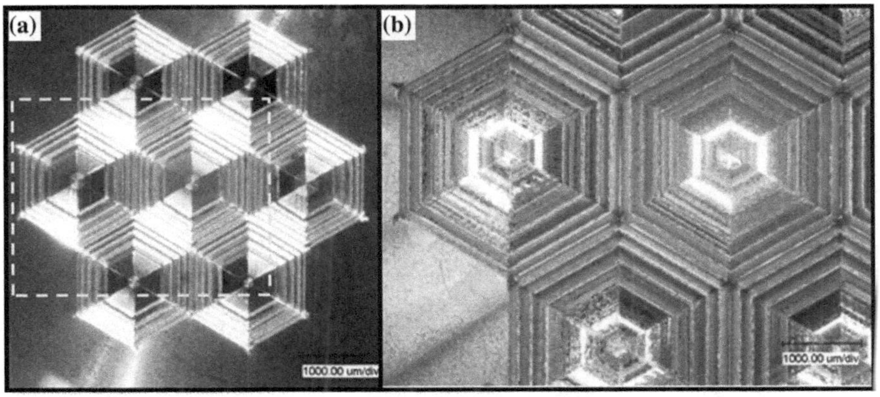


**Fig. 7.13** Photographic images of a machined hexagonal Fresnel lens array; **a** full view and **b** an enlarged view for the selected zone (*dashed box*). Figure reproduced from [9]

using an Olympus LEXT OLS4000 3D Measuring Laser Microscope with a confocal optical system. 3D measured profile data were further post-processed using MATLAB for surface characterization.

Figure 7.14a and e show the 3D contour measurements of the central and offset hexagonal Fresnel lenses, respectively. As illustrated in Fig. 7.14b–d and f–h, the measured $E_\rho$ for the selected zones of the central and offset lenses, respectively, are able to achieve lesser than the required $E_\rho$ of 0.1 μm. On the other hand, Fig. 7.15a and e present the selected areas of 3D contour measurements of the central and offset hexagonal Fresnel lenses, respectively, for the evaluation of sagittal errors $h_{err}$. From Fig. 7.15b–d and f–g, the measured $h_{err}$ of respectively areas are also found to be lesser than the required $h_{tol}$ of 0.1 μm.

Thus, these results further validate that the selected machining parameter, the feedrate $\Delta\rho$, for cutting the hexagonal Fresnel lens array is achieving the cutting residual error $E_\rho$ lesser than 0.1 μm. On top of these, these Sagitta error $h_{err}$ results also further validate that the selected second parameter $\Delta t$ for machining hexagonal Fresnel lens array are also achieving the accuracies lesser than the $h_{tol}$ of 0.1 μm.

Lastly, these experimental results have validated the credibility of the proposed Guilloche machining technique to fabricate accurate hexagonal Fresnel lens array in a single process. In additions, this can be only achieved with the implementation of the proposed approaches, namely cutting residual error and Sagitta error analyses.

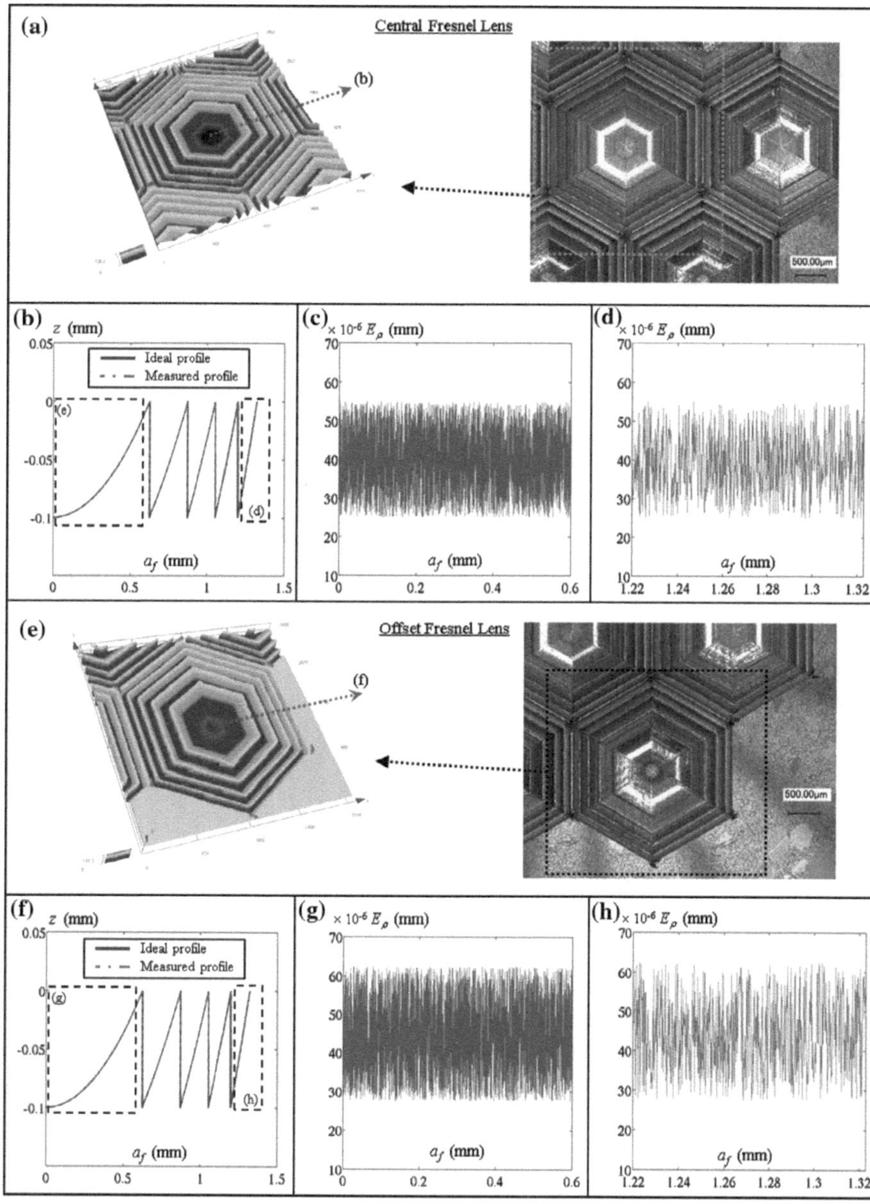


**Fig. 7.14** Cutting residual error of machined hexagonal Fresnel lenses

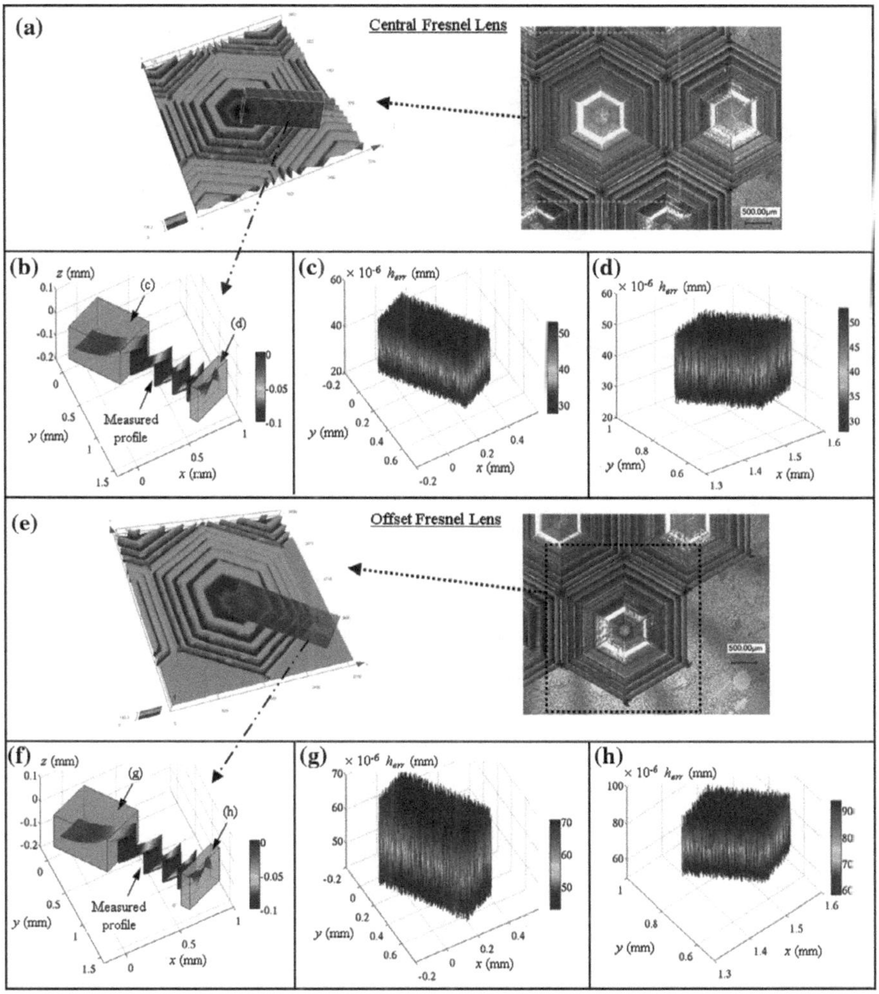


**Fig. 7.15** Measured Sagitta errors of machined hexagonal Fresnel lenses

## 7.3 Case Study 2: Multiple-Compound Eye Surface Design-B

### 7.3.1 *Experimental Validations*

In this case study, a multiple-compound eye surface, as illustrated in Fig. 7.16. has been machined by hybrid fast tool/slow slide servo (FTS/SSS) diamond turning process. Table 7.3 describes the selected experimental conditions for the cutting experiments. A rapidly solidified aluminum alloy RSA-6061 has been employed as a workpiece material for the cutting experiments. RSA-6061 is an ultra-fine grain

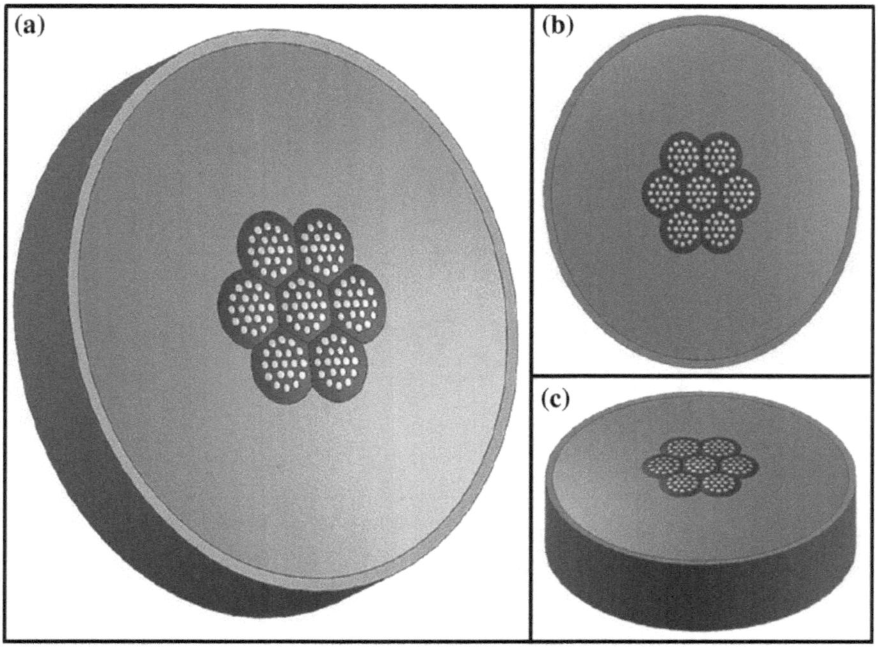


**Fig. 7.16** CAD model for multiple-compound eye Design-B, **a** 3D view, **b** top view, and **c** slanted view

**Table 7.3** Cutting conditions for case study 2

| | |
|---|---|
| Feedrate | 0.005 mm/rev (Rough cuts)<br>0.001 mm/rev (Finish cuts) |
| Spindle speed | 50 rev/min (Both rough and finish cuts) |
| Depth of cut | 0.010 mm (Rough cuts)<br>0.001–0.005 mm (Finish cuts) |
| Workpiece | RSA-6061 |
| Tool #1 for rough cuts | Mono-Crystal Diamond tool nose radius = 0.2 mm<br>$\gamma_{tool} = 0°$, $\alpha_{tool} = 5°$, $\beta_{tool} = 60°$ |
| Tool #2 for finish cuts | Mono-Crystal Diamond tool nose radius = 0.01 mm<br>$\gamma_{tool} = 0°$, $\alpha_{tool} = 30°$, $\beta_{tool} = 30°$ |
| Targeted $PV_{err}$ | 2.0 μm (Rough); 1.0 μm (Finish) |

aluminum alloy AA-6061, which is widely used for making optical inserts owning to its excellent nanometric surface finishing after diamond turning process. Furthermore, its fine microstructures by a superfast solidification in the melt-spinning process gives excellent mechanical and physical properties [13, 14].

At the same time, two different types of diamond tool geometries have also been employed for roughing and finishing processes. In the roughing process, a diamond tool having a nose radius of 0.2 mm, flank clearance of 7° and wedge angle of 60°

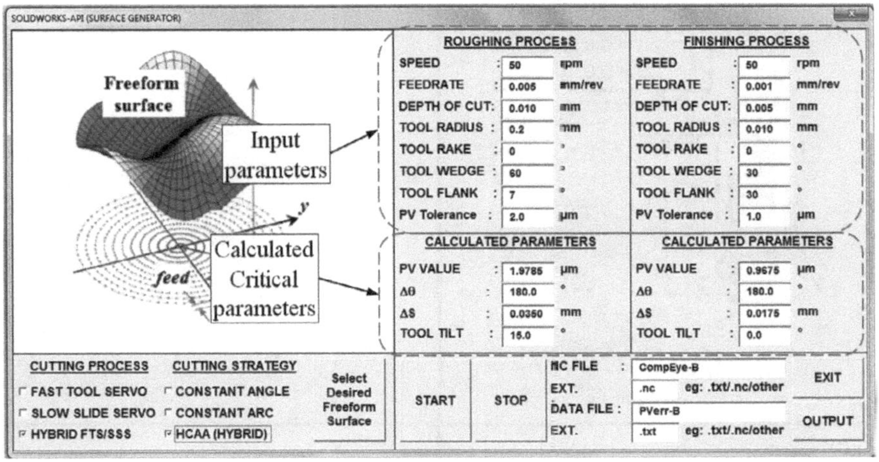


**Fig. 7.17** Screenshot image for the calculated critical parameters by the developed integrated system. Figure reproduced from [3]

is employed with the cutting feedrate of 0.005 mm/rev and depth of cuts of 10 μm. As for the finishing process, a diamond tool with nose radius of 10 μm, flank clearance of 30° and wedge angle of 30° is employed with the cutting feedrate of 0.001 mm/rev and depth of cuts of 5 μm.

Figure 7.17 presents a screenshot image of user-interface with the generated critical cutting parameters and tool geometrical angles by the developed system. It can be seen that the critical cutting parameters, namely constant angle $\Delta\theta$, constant arc-length $\Delta S$ and transition radius $r_{trans}$, have been pre-defined for both roughing and finishing cutting experiments to achieve accurate freeform contours. Furthermore, the critical tool geometrical angles, namely $\gamma_{cr}$, $\alpha_{cr}$ and $\beta_{cr}$ are also been evaluated to avoid any unnecessary tool interferences. The analytical details of these critical parameters can be extracted from the output files of the system and are presented in the next sections.

#### 7.3.1.1 Critical Cutting Parameters for HCAA Cutting Strategy

Figure 7.18 illustrates a replication of output data from the developed system for the critical cutting parameters in the hybrid constant-arc and constant-angle (HCAA) cutting strategy. It can be seen that the constant angle cutting strategy has been selected for the outer and inner regions, and the middle region's cutting strategy is the constant arc. The transition radii, $r_{trans1}$ and $r_{trans2}$ between these regions are located at 3.9865 and 0.2782 mm, respectively. In the constant arc cutting strategy, the critical value of $\Delta S$ for the finishing and roughing cuts, as illustrated in Fig. 7.18a, should not exceed 0.0175 and 0.035 mm, respectively. The calculated $PV_{err}$ values of middle region for the finishing and roughing cuts are

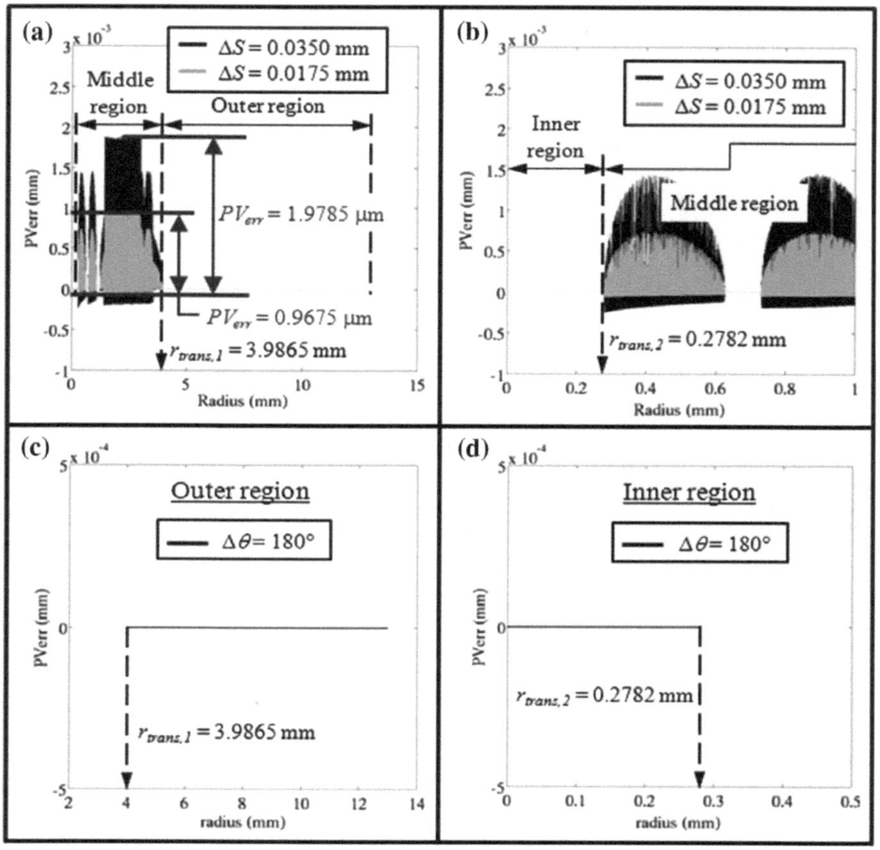


**Fig. 7.18** Replication of output data for the details of critical cutting parameters from developed system. Figure reproduced from [3]

0.9675 μm (<1.0 μm) and 1.9785 μm (<2.0 μm), respectively, which are also lesser than the targeted $PV_{err}$. On the other hand, the critical $\Delta\theta$ for both finishing and roughing cuts, as shown in Fig. 7.18c and d, are the same value of 180°. The calculated $PV_{err}$ values of outer and inner regions are found to be near zero for both finishing and roughing cuts.

In summary, the cutting process begins with a constant-angle cutting strategy from outermost radius until the tool reaches $r_{trans1}$ of 3.9865 mm. Then, the cutting process continues with constant-arc cutting strategy for the middle region. When the tool reaches the next $r_{trans2}$ of 0.2782 mm, the cutting process reverts back to constant-angle cutting strategy for the inner region. This entire cycle shall be repeated for each cut until the whole cutting process completes.

#### 7.3.1.2 Critical Tool Geometrical Angles

Figure 7.19 represents the replication of the critical tool geometrical angles from output data of the developed system. The critical values of tool rake $\gamma_{cr}$ and front clearance $\alpha_{cr}$ angles are 70° and 20°, as illustrated in Fig. 7.19a and b, respectively, and the critical included angle $\beta_{cr}$, as shown in Fig. 7.19c, is about 135°. With the respect to above results, it can be seen that Tool #2 would be free of tool interferences but not for Tool #1. There would be a frank interference in the case of Tool #1 as $\alpha_{tool}$ of 5° is lesser than the critical one ($\alpha_{cr}$ = 20°). Thus, Tool #1 is required to be tilted at $\lambda_{cr}$ of 15°, as illustrated in Fig. 7.20, overcoming the frank interference with an additional spacer.

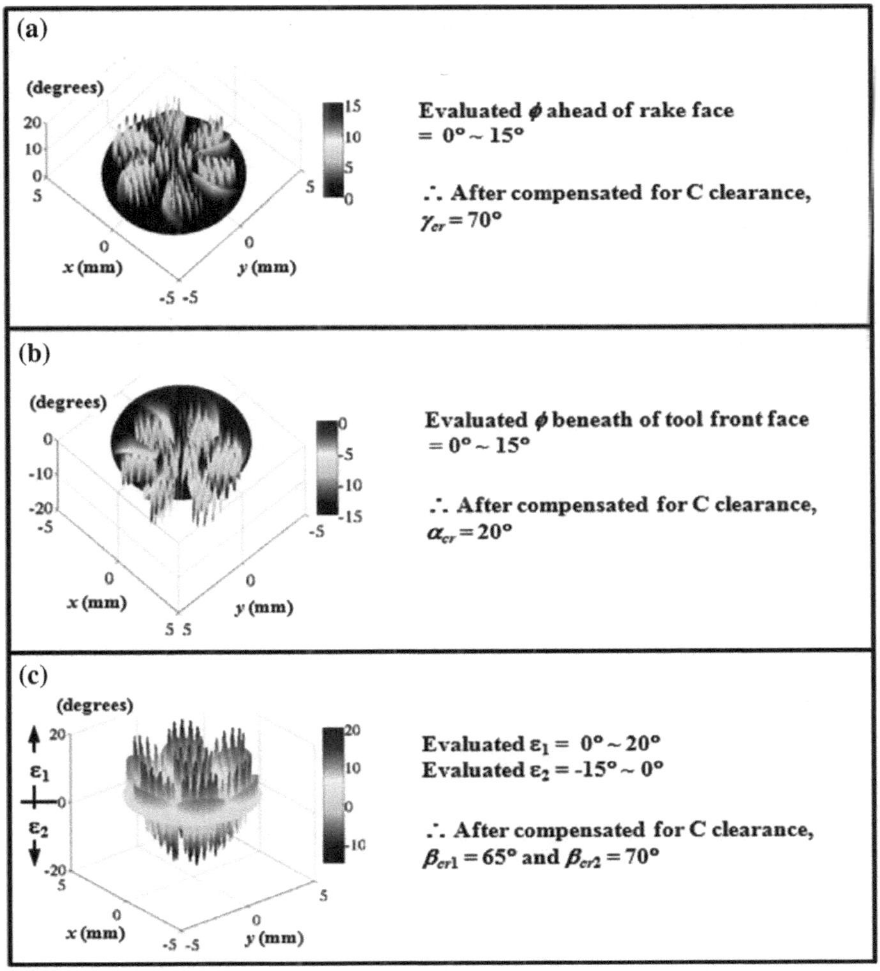


**Fig. 7.19** Replication of output data for the details of critical tool angles from developed system. Figure reproduced from [3]

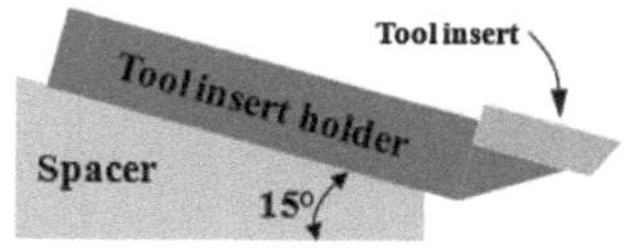


Fig. 7.20 Schematic setup for tilting Tool #1

#### 7.3.1.3 Geometrical Splitting for Hybrid FTS/SSS Process

Figure 7.21 replicates the critical geometrical frequencies from output data of the developed system. It demonstrates that the generated tool trajectories for hybrid freeform surface has two notable frequencies of 6.098 and 47.03 Hz, as illustrated in Fig. 7.21a. Hence, the generated tool trajectory can be further split into two types of trajectories, namely low-order and high-order, as shown in Fig. 7.21b and c, respectively. Thus, the low-order tool trajectory having a frequency property of 6.245 Hz and a sag height of 0.3 mm would be handled by SSS process. On the other hand, the other high-order one having a frequency property of 46.48 Hz with a sag height of 0.02 mm would be handled by FTS process.

### 7.3.2 *Cutting Experiments and Results*

Several cutting experiments have been performed by employing a hybrid fast tool/slow slide servo (FTS/SSS) diamond turning process. hybrid constant-angle

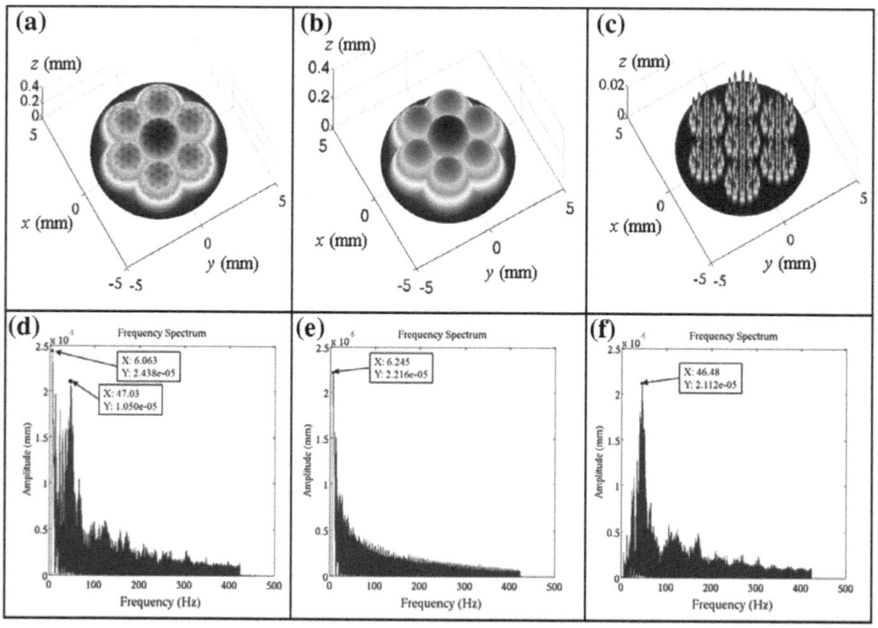


Fig. 7.21 Geometrical splitting of freeform features; **a–c** 3D profiles and **d–f** calculated frequency spectrums for the original, low-order (SSS) and high-order (FTS) tool trajectories, respectively

and constant-arc (HCAA) cutting strategy. The cutting conditions have been chosen based on the above results to fabricate the multiple-compound eye surface. Figure 7.22 shows that the generated spiral tool trajectory points have been successfully overlayed on the CAD surface by the developed SolidWorks-API system. These tool trajectory points were further post-processed into NC codes for the cutting process

Figure 7.23 illustrates the photographic images for the successful machining of the multiple compound eye surface. The machined surface were measured using a 3D measuring laser microscope with a confocal optical system (Olympus LEXT OLS4000). 3D measured surface data were further characterized using MATLAB software for the surface error.

Figures 7.24 illustrates the replicated of 3D contour measurements for the central and offset compound eye surfaces. Figure 7.24b–c and d–e show that the measured contour errors for the selected regions of the central and offset compound eye surfaces, respectively, were found to be lesser than the targeted $PV_{tol}$ of 1.0 μm. Hence, these results are validating the proper selection of critical cutting parameters, $\Delta S$ and $\Delta\theta$, for the machining of the freeform surface using hybrid constant-arc and constant-angle cutting strategy.

In additions, no tool interference/rubbing marks are detected on the machined surface. Therefore, these experimental results have validated the credibility of the proposed methodologies, hybrid Hilbert and fast Fourier transformations HT/FFT, and tool geometrical optimization, for incorporating into the integrated SoildWorks-API system to fabricate hybrid freeform surface accurately.

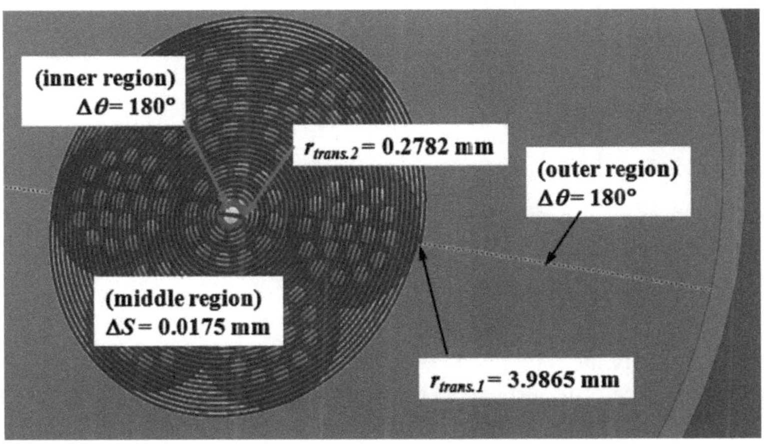


**Fig. 7.22** A successful generation of tool trajectories for HCAA cutting strategy, which are mapped onto the surface of the CAD model. Figure reproduced from [3]

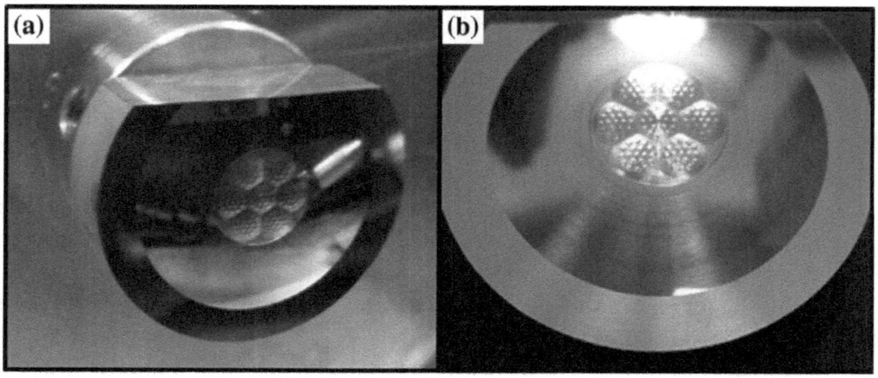


**Fig. 7.23** Photographic images of fabricated multiple compound eye surface; **a** fabricated workpiece mounted on a chuck, and **b** top slanted view. Figure reproduced from [3]

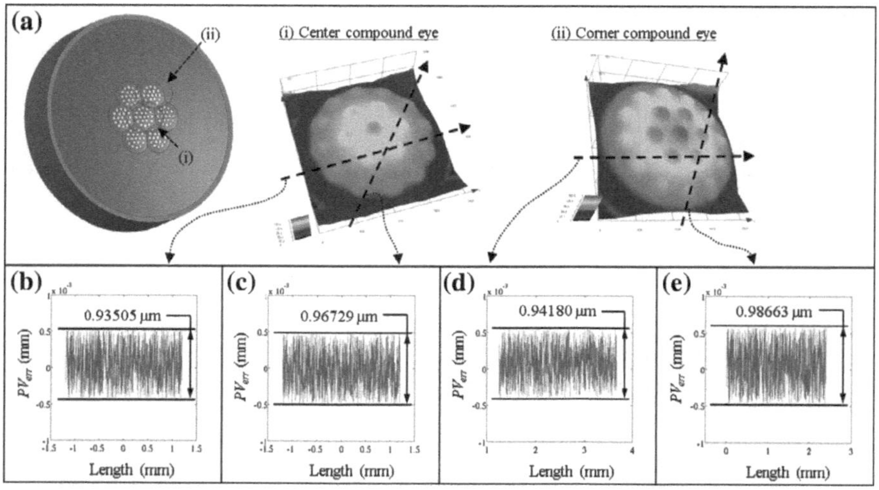


**Fig. 7.24** Contour error measurements of the central compound eye surface

## 7.4 Concluding Remarks

In the present study, an integrated CAD/CAM system has been developed to fabricate complex freeform surface accurately using multiple-axis diamond turning processes. The proposed methodologies not only replaces the needs of expensive specialized CAM software, but also provides an attractive solution with an integration of Visual Basic application programming interface (API) into SolidWorks for accurate and optimized surface generation of hybrid freeform surfaces. The conclusions are been drawn as follows:

i. Two case studies explaining the implementation of the developed integrated system is presented. The profile accuracy requirements for the hybrid freeform surface have been met. The contour evaluation has demonstrated that the profile errors were lesser than the targeted requirements.

ii. This can be only achieved with the implementation of several analytical approaches for accurate freeform surfaces. These approaches evaluate analytically for the cutting residual, Sagitta and cutting linearization errors.
iii. The tool interference/rubbing marks are also eliminated with an aid of tool geometrical optimization approach.
iv. In addition, the hybrid approach of Hilbert transformation (HT) and fast Fourier transformation (FFT) performs a proper segregation of freeform features in order to be fabricated by FTS and SSS processes simultaneously.

This study provides an essential contribution towards the improvement of CAD/CAM supports for multiple-axis ultraprecision machining of complex hybrid freeform surfaces.

## References


1. Yu DP, Wong YS, Hong GS. Optimal selection of machining parameters for fast tool servo diamond turning. Int J Adv Manuf Technol. 2011;57:85–99.
2. Hsu WY, Chen FZ, Kuo CH, Cheng YC, Lin WL. Fabrication of micro aspherical lens array using fast tool servo with tool path compensation. In: ASPE proceedings, annual meetings; 2008.
3. Neo DWK, Kumar AS, Rahman M. CAx technologies for hybrid fast tool/slow slide servo diamond turning of freeform surface. Proc Inst Mech Eng Part B J Eng Manuf. 2016;230(8):1465–79.
4. Cooley JW, Lewis PAW, Welch PD. The fast fourier transform and its applications. IEEE Trans Educ. 1969;12(1):27–34.
5. Noda I. Determination of two-dimensional correlation spectra using the Hilbert transform. Appl Spectro. 2000;54(7):994–9.
6. Le Van Quyen M, Foucher J, Lachaux J, Rodriguez E, Lutz A, Martinerie J, Varela FJ. Comparison of Hilbert transform and wavelet methods for the analysis of neuronal synchrony. J Neurosci Methods. 2001;111(2):83–9.
7. Zadeh LA, Desoer CA. Linear system theory. New York: Mc Graw Hill; 1969.
8. King FW, Smethells GJ, Helleloid GT, Pelzl PJ. Numerical evaluation of Hilbert transforms for oscillatory functions: a convergence accelerator approach. Comput Phys Commun. 2002;145(2):256–66.
9. Neo DWK, Kumar AS, Rahman M. Automated Guilloche machining technique for the fabrication of polygonal Fresnel lens array. Precis Eng. 2015;41:55–62.
10. Alda J, González FJ. Polygonal Fresnel zone plates. J Opt A Pure Appl Opt. 2009;11(8):085707. doi:10.1088/1464-4258/11/8/085707.
11. Gao N, Xie C. Experimental demonstration of free-space optical vortex transmutation with polygonal lenses. Opt Lett. 2012;37(15):3255–7. doi:10.1364/OL.37.003255.
12. Burgess EL, Marchi DL, Walter H. A novel square fresnel lens design for illuminating circular solar cells. Opt Eng. 1978;17(3):299–300. doi:10.1117/12.7972232.
13. Karakösea E, Keskin M. Structural investigations of mechanical properties of Al based rapidly solidified alloys. Mater Des. 2011;32(10):4970–9.
14. Abou-El-Hossein K, Olufayo O, Mkoko Z. Diamond tool wear during ultra-high precision machining of rapidly solidified aluminium RSA 905. Wear. 2013;302(1–2):1105–12.

# Chapter 8
# Conclusions and Recommended Future Works

## 8.1 Major Contributions

In this study, an attempt of integrating the multiple-axis diamond turning processes with the proposed methodologies into the CAD/CAM system was made. Hence, there are four major contributions for optimizing the ultraprecision manufacturing of hybrid freeform surfaces and are summarized as follows:

i. Developed methodologies for hybrid FTS/SSS diamond turning

   - Hybrid fast tool/slow slide servo (FTS/SSS) diamond turning was developed by incorporating both FTS and SSS techniques to optimize the fabrication process of hybrid freeform surfaces and address the limited range of FTS stroke length and the low bandwidth in the SSS system.
   - Hybrid fast Fourier and Hilbert transformation (FFT/HT) method has been developed for the segregation of freeform features based on their frequency domains.

ii. Developed novel multiple-axis surface generation for complex freeform surfaces

   - A novel multiple-axis surface generation methodology has been developed to fabricate a complex freeform surface, such as polygonal Fresnel lens array. It addresses the difficulties of fabricating such surfaces in a single setup.

iii. Developed novel surface analytical model for cutting linearization errors

   - The accuracy of machined freeform surface can be pre-evaluated with the derived novel surface analytical model before machining.
   - The model evaluates the cutting linearization errors along the spiral tool trajectory of fast tool/slow slide servo diamond turning process.

 
D.W.K. Neo, *Ultraprecision Machining of Hybrid Freeform Surfaces Using Multiple-Axis Diamond Turning*, Springer Theses,
DOI 10.1007/978-981-10-4083-2_8

- The number of cutting points can be optimized for achieving the targeted profile accuracy.

iv. Developed integrated CAD/CAM software package for machining freeform surfaces

- A comprehensive, integrated CAD/CAM software solution for multiple-axis diamond turning has been developed for planning and conducting the manufacture of hybrid freeform surface and to make available open interfaces for the different adaption technologies.
- With the implementation of the integrated system, the profile accuracy requirements can be met. Moreover, a proper segregation technique of freeform features has also been established for fabrication of hybrid freeform surface using hybrid FTS/SSS process.
- Furthermore, the tool interference/rubbing marks are also eliminated with an aid of tool geometrical optimization approach.
- These provide an essential contribution towards the improvement of CAD/CAM supports for ultraprecision machining of complex hybrid freeform surfaces.

## 8.2 Recommended Future Works

i. Discrete arc-length cutting strategy for better optimization of cutting linearization errors
Although the hybrid constant-arc and constant-angle (HCAA) cutting strategy offers the best optimization results in this study, there is still room for improving the optimization of controlling cutting linearization errors ($PV_{err}$). From the experimental studies, it can be concluded that $PV_{err}$ is dependant of the curvatures of freeform surface and is not necessary dependant on the arc-lengths.
Hence, this promotes another method of controlling the arc-lengths in which the arc-lengths should be maximized to the longest possible distance as long as the $PV_{err}$ does not exceed the required profile accuracy tolerance. Thus, there will be several discrete arc-lengths throughout the entire freeform surface. This discrete arc-length cutting strategy not only optimizes the number of controlled points, but also improves the overall efficiency of manufacturing duration of freeform surfaces.

ii. Mechanism of chip formation in freeform cutting
Ultraprecision cutting is one of the dominant approaches to obtain intricate features and high surface finish. Therefore, it is significant of understanding the material removal mechanism in nanometric scale, which helps to achieve a better surface finish and increase the process efficiency as well as to obtain high

economic value. The mechanism of chip formation in the microcutting is greatly influenced by the tool rake angles.

Up to date, all works on chip formation in the ultraprecision cutting have been only conducted with rake angles individually. However, the effective rake angle changes all the time during the diamond turning of freeform surface, even with a zero rake diamond tool. The effective rake angle in a diamond turning of freeform surfaces can be varied from positive to negative values depending on the slopes of surface. This variation of effective rake angles would also cause the variation of chip formation which would behave as elastic-plastic deformation.

However, there is no literature studying the optical effect of freeform diamond turning under the influence of varying effective rake angles. Therefore, it is worth to consider studying the chip formation under the influence of effective tool rake angles in ultraprecision machining of freeform optical surfaces.

iii. Sophisticated compensation method for the surface analytical model

The developed surface analytical model is purely geometrical. It is well known that the surface generated during machining is strongly dependant of material properties, and cutting mechanisms (e.g. side flow, indentation, etc.). These mechanical effects would affect the final machined surface quality and accuracy.

One such mechanical effect is the variation of chip formation (elastic-plastic deformation) due to the varying rake angles on the freeform curvatures. Thus, the consideration for the varying rake angles should be taken into an account in improvising the surface analytical model with better predictions of good surface quality and accuracy.

iv. Hybrid AGMT and FTS/SSS process

Presently, AGMT and FTS/SSS processes are two separate and distinct ones, which are only able to fabricate a limited range of hybrid freeform surfaces, depending on their capabilities. This has been already explained in this dissertation that this is due to the loss of symmetry in the complex freeform surface.

Hence, a hybrid of both processes would enhance their capabilities in addressing the difficulties in the machining of more complex hybrid freeform surfaces, i.e. an array of sinusoidal Fresnel lenses.

# Curriculum Vitae

## Personal Data

| | |
|---|---|
| Full Name: | Dennis Wee Keong, Neo |
| Date of birth: | 25 July 1977, Singapore |
| Professional Address: | Dept. of Mech. Eng., National University of Singapore, 9 Engineering Drive 1, Singapore 117576 |
| Email 1: | mpenwk@nus.edu.sg |
| Email 2: | dennis.sm07@gmail.com |


## Education

National University of Singapore, Singapore

- Graduate research with A. Senthil Kumar and Mustafizur Rahman, Jan 2011 to Dec 2014
- Ph.D., Mechanical Engineering
- Thesis: "Ultraprecision machining of hybrid freeform surfaces using multiple-axis diamond turning"

National University of Singapore, Singapore

- B.Tech., Mechanical Engineering, Jan 2005 to Dec 2009
- Dean's list for outstanding performance in Semester I, 2009/2010



D.W.K. Neo, *Ultraprecision Machining of Hybrid Freeform Surfaces Using Multiple-Axis Diamond Turning*, Springer Theses,
DOI 10.1007/978-981-10-4083-2

**Award/Honors**

1. HEIDENHAIN Scholarship from 14th EUSPEN Conference, 2014
2. Top downloaded paper in Proceedings of the Institution of Mechanical Engineers, Part B: Journal of Engineering Manufacture, 2013
3. NUS PhD scholarship, Jan 2011
4. Dean's list award for excellent academic performance in the year 2009/2010

**Work Experience**

National University of Singapore, Singapore—Aug 2015 to Present

Research Fellow

- Leads a group of research engineers/associates and PhD students in several research areas of deep hole drilling project
- Provides advice to achieve critical scientific and technological breakthroughs
- Plan and optimize the research process to deliver sustainability performance
- Prepares proposals for gaining more research and industrial funding

Mikrotools Pte Ltd, Singapore—Jan 2015 to Aug 2015

Senior Process Engineer

- Provides professional advices on the ultraprecision machining of advanced freeform surfaces
- Provides professional training to users in understanding ultraprecision diamond turning machine
- Provides professional solutions and consultancy for ultraprecision machining services

Joo Huat Engineering Pte Ltd, Singapore—Jul 2003 to Sep 2010

Quality Assurance Engineer

- Determines all the daily quality activities are being performed according to all the quality requirements.
- Prepares First Article Inspections (FAI) for the Customer's approval and approves the first piece setups inspection for the production running.
- Involves in the preparation for the Quality Assurance Plans (QAP) for the projects accordance to the Customer's specific requirements.
- Prepares the process plans and designs the fixtures for new production parts/methods

Winmech Precision Pte Ltd, Singapore—1999 to 2002

CNC Machinist

- In-charge of first piece machining setups and approves the first piece setups inspections for the production running smoothly.
- Closely supervises a group of CNC operators to ensure a smooth production running.

**Research Interests**

Micro/nano machining and manufacturing of freeform surfaces, fast tool/slow slide servo, multiple-axis ultraprecision machining techniques, process optimization, surface error analysis, deep hole drilling, and large format machining.

**Professional Development**

Technology skills: Microsoft office, Unigraphics, AutoCAD, SolidWorks, MatLab

Training: Internal Auditor, ISO9001/AS9100

Languages: English, Chinese

**Publications and Pre-prints**
**Citations: 49, h-index: 4, i10-index: 2**

*Book Chapter*

1. A. Senthil Kumar, M.R. Aravind Raghavendra, W.K. Neo, M. Rahman. Fast and fine tool servo for ultraprecision machining in comprehensive materials processing. In: Rahman, M. editor. Advanced machining technologies, vol. 11. Elsevier Ltd; 2014. pp 61–88, 10.1016/B978-0-08-096532-1.01104-3.

*Journals*

1. D.W.K. Neo, A.S. Kumar and M. Rahman. CAx technologies for hybrid fast tool/slow slide servo diamond turning of freeform surface. In: Proceedings of the institution of mechanical engineers, part B, J. Engineering manufacture, 2016; In-Press.
2. D.W.K. Neo, A.S. Kumar and M. Rahman. Automated Guilloche machining technique for the fabrication of polygonal Fresnel lens array. Precision Eng. 2015; 41:55–62.
3. D.W.K. Neo, A.S. Kumar and M. Rahman. A novel surface analytical model for cutting linearization error in fast tool/slow slide servo diamond turning. Precision Eng. 2014; 38(4):849–860.
4. X. Mu, D.W.K. Neo, H. Yu, A.S. Kumar and F.S. Chau. Polygonal pyramidal reflector-based micromachined micro-scanners for bioimaging. J. Micro/Nanolith. MEMS MOEMS (Jan–Mar 2014); 13(1):011109.
5. X. Mu, G. Zhou, H. Yu, J.M.L. Tsai, W.K. Neo, A.S. Kumar and F.S. Chau. Electrostatic MEMS resonating micro polygonal scanner for circumferential endoscopic bio-imaging. In: Proceeding of SPIE 8616, MOEMS and miniaturized systems XII, 861606 (2013).

6. X. Mu, G. Zhou, H. Yu, J.M.L. Tsai, W.K. Neo, A.S. Kumar and F.S. Chau. Electrostatic micromachined resonating micro-scanner for circumferential endoscopic bio-imaging. IEEE Photonics Technol Lett. 2013; 25(8):749–752.
7. X. Mu, G. Zhou, H. Yu, J.M.L. Tsai, D.W.K. Neo, A. S. Kumar and F.S. Chau. MEMS Electrostatic Double T-shaped Spring Mechanism based MEMS Device for Circumferential Scanning. IEEE J. Microelectromechanical Syst. 2013; 22 (5):1147–1157.
8. W.K. Neo, A. S. Kumar and M. Rahman. A novel method for layered tool path generation in the fast tool servo diamond turning of non-circular microstructural surfaces. In: Proceedings of the institution of mechanical engineers, part b, j. engineering manufacture. 2013. 227(2):210–219. (Top most downloaded paper).
9. W.K. Neo, A.S. Kumar and M. Rahman. A review on the current research trends in ductile regime machining. Int J Adv Manuf Technol. 2012; 63(5):465–480.

*Conference Papers*

1. W.K. Neo, A.S. Kumar and M. Rahman. Novel micro-grooving technique for machining of novel chevron sharkskin riblets on flat surface. In: 14th international conference and exhibition on european society for precision engineering & nanotechnology (EUSPEN 2014), Dubrovnik, Croatia, 2nd to 6th June 2014.
2. W.K. Neo, A.S. Kumar and M. Rahman. Profile Accuracy Model for Hybrid Constant-Arc and Constant-Angle Method in the FTS/STS Diamond Turning. In: 9th international conference on micro manufacturing (ICOMM 2014), Nanyang Technological University, Singapore, 25th to 28th March 2014.
3. W.K. Neo, A.S. Kumar and M. Rahman. A novel method for profile error analysis of freeform surfaces in FTS/STS diamond turning. In: 5th international conference of asian society for precision engineering and nanotechnology (ASPEN2013), Taipei, Taiwan, 12 to 15 Nov 2013.
4. W.K. Neo, A.S. Kumar and M. Rahman, Offset diamond turning technique for machining of Fresnel lens arrays. In: 10th international conference on multi-materials micro manufacture (4M2013), San Sebastian, Spain, 8th to 10th Oct 2013.
5. W.K. Neo, R. Huang, A.S. Kumar, and M. Rahman. Novel Machining Technique for Surface Patterning by Diamond Turning. In: 8th international conference on micro manufacturing (ICOMM 2013), Victoria, BC, Canada, 25th to 28th March 2013
6. X. Mu, G. Zhou, H. Yu, J.M.L. Tsai, W.K. Neo, A.S. Kumar and F.S. Chau. Electrostatic MEMS resonating micro polygonal scanner for circumferential endoscopic bio-imaging. In: Proceeding SPIE 8616, MOEMS and miniaturized systems XII, 861606 (Photonic West 2013), San Francisco, California, United States, 4th to 6th February 2013, 10.1117/12.2003972.
7. R. Huang, W.K. Neo, A.S. Kumar, K. Liu and M. Rahman. An Innovative Method of Machining Freeform Feature Based on Digital Image by Fast Tool Servo Diamond Turning. In: 4th international 25th all india manufacturing

technology design and research conference (AIMTDR 2012), Kolkata, West Bengal, India, 14th to 16th Dec 2012.

8. M. Rahman, A.S. Kumar and W.K. Neo. Current Status and Future Challenges in Ultraprecision Machining of Freeform Optical Surfaces. In: Keynote paper, 3rd asia pacific conference on optics manufacture (APCOM 2012), Changchun, China, 26th to 28th Aug 2012.

## Academic References

| | |
|---|---|
| A/Prof. A. Senthil Kumar<br>Department of Mechanical Engineering<br>University of Singapore<br>9 Engineering Drive, EA Building<br>EA-05-25<br>Singapore 117576<br>(+65) 65166800<br>mpeask@nus.edu.sg | Prof. Mustafizur Rahman<br>Department of Mechanical Engineering<br>University of Singapore<br>9 Engineering Drive, EA Building<br>EA-05-11<br>Singapore 117576<br>(+65) 65162168<br>mpemusta@nus.edu.sg |

FSC
www.fsc.org
MIX
Papier | Fördert
gute Waldnutzung
FSC® C083411

Kolibri 360 GmbH
Ferdinand-Jühlke-Straße 7
99095 Erfurt, Deutschland
produktsicherheit@kolibri360.de